TRANSACTIONS

OF THE

AMERICAN PHILOSOPHICAL SOCIETY

HELD AT PHILADELPHIA

FOR PROMOTING USEFUL KNOWLEDGE

NEW SERIES—VOLUME 50, PART 2

1960

STUDIES IN BYZANTINE ASTRONOMICAL TERMINOLOGY

O. NEUGEBAUER

Professor of the History of Mathematics,
Brown University

THE AMERICAN PHILOSOPHICAL SOCIETY

INDEPENDENCE SQUARE

PHILADELPHIA 6

January, 1960

To the memory of my teacher of Latin and Greek

R. WIMMERER

late Professor at the k. k. I. Staatsgymnasium in Graz

who never despaired of educating even his most hopeless pupils

Library of Congress Catalogue
Card No. 59-00000

STUDIES IN BYZANTINE ASTRONOMICAL TERMINOLOGY

O. NEUGEBAUER

CONTENTS

INTRODUCTION

To the best of my knowledge the first and also last study of mathematical astronomy of the Byzantine period consists in a short chapter of the *Astronomia Philolaica* by Bullialdus (Paris, 1645). Peripheral subjects have been treated more often; it may suffice to recall Honigmann's utilization [1] of lists of " famous cities," given customarily with astronomical tables, or A. Delatte's publication of treatises on the astrolabe.[2] There exists a very extensive literature on the Easter computus and several studies of elementary treatises, e.g. Heiberg's publication of an anonymous " Quadrivium." [3] Nearest to Bullialdus' work comes Usener's *De Stephano Alexandrino* (1888), in particular Stephanus' treatise on Theon's Handy Tables.[4]

With the scarcity of these attempts must be contrasted the enormous wealth of astronomical manuscripts which are scattered over the European libraries. Since it took, under the leadership of Boll and Cumont, about fifty years to complete the twelve volumes of the Catalogus Codicum Astrologorum Graecorum one cannot entertain under present conditions any hopes of accomplishing a similar survey for the astronomical tables and treatises. Needless to say, a critical edition of thousands of folios of astronomical works is out of the question. Furthermore, even a survey would only be of value to the history of astronomy if it would display the basic numerical material as well as the theoretical techniques used by the individual sources. This, however, would require a far deeper penetration into the contents of such texts, than was needed for the astrological material, particularly with Bouché-Leclercq's superb work on Greek astrology at one's disposal.

[1] *Die sieben Klimata*, Heidelberg, 1929.

[2] *Anecdota Atheniensia* II, Paris, 1939.

[3] Copenhagen, 1929.

[4] Usener, Kl. Schr. III, p. 295 to 319. *Cf.* also his excellent study " Ad historiam astronomiae symbola " (1876) though here the emphasis lies on chronology and calendariography.—Tannery's Sciences exactes chez les Byzantines (Mémoires Scientifiques IV, Paris, 1920) contains very little on mathematical astronomy.

The only way left open seems, therefore, to be the investigation in detail of some specific documents which at the outset promise to be of interest. To find such texts causes no difficulties. Heiberg in his *Byzantinische Analecten* [5] gave a short description of Cod. Vat. Gr. 1058 containing among others "ein anonymes astronomisches Werk in vielen Kapiteln" (beginning with a method for finding the solar longitude and a reference to Persian years) and another "grösseres astronomisches Werk . . . von Σαντζαρῆ"—obviously meaning the Sanjarī zīj.[6] However short, Heiberg's excerpts contain several unknown terms [7] and I, therefore, decided to make an attempt to clarify at least the terminology of this section of the text.[8] I am grateful to the courtesy of the Biblioteca Apostolica Vaticana for permission to use a film of this codex and of related manuscripts and to publish my findings.

What follows is a simple list of technical terms, their meaning being determined by their usage in Cod. Vat. gr. 1058, fol. 261 to 459.[9] In the more involved cases, the appendices will furnish the proofs for my translations from larger context. Many of the terms are translations or even loan words from Persian and Arabic. For these problems I had the help of my friend and colleague, Professor E. S. Kennedy, whose experience with a large body of Islamic treatises and tables (*zījes*)[10] provided most useful information about many of the sources from which this class of Byzantine astronomical texts is derived.

Beside treatises which were either direct translations of Persian works or which were at least greatly influenced by Islamic sources, we find also the direct Greek tradition alive, e.g., in tables ultimately based on Ptolemy and Theon. In the middle of Vat. gr. 1058, we find even a short section [11] of purely Greek origin. In manuscripts of this composite type each individual section poses problems of its own and I have made no attempt to solve them all, as would have been necessary for an edition of the text. The references usually give no more than one characteristic occurrence, sufficient to establish the technical meaning from the proper context. It was not my intention to give a complete glossary, but only to collect all terms whose meaning was not obvious from general Greek usage or from their occurrence in the Almagest.

I have followed the manuscript rather closely and adopted its disregard for subscript iotas, capitals in proper names, etc. Not being a philologist, I feel perfectly at ease in the company of such defects. All that I can claim is a reasonably complete understanding of the subject matter of these treatises. On this basis I hope that this study, however fragmentary, may be found helpful for subsequent investigations of mediaeval astronomy.

NUMERICAL NOTATION

The majority of numbers in astronomical tables are written in the system well known from the Almagest: integers, in particular degrees from 1 to 360, are written with the Greek alphabetic numerals displayed in table 1.

TABLE 1

	1	10	100	1000
1	α	ι	ρ	,α
2	β	κ	σ	,β
3	γ	λ	τ	,γ
4	δ	μ	υ	
5	ε	ν	φ	etc.
6	ϛ	ξ	χ	
7	ζ	ο	ψ	
8	η	π	ω	
9	θ	ϟ	ϡ	

Fractions, however, are always written sexagesimally, of course again using the Greek alphabetic numerals for the single sexagesimal digits from 1 to 59. We transcribe [12]

$\overline{\tau\ \nu\delta\ \kappa\beta\ \alpha\ \lambda\varsigma\ \nu\alpha}$ by 354;22,1,36,51.

The letters are in many cases written in so close a sequence that the separation into decimal or sexagesimal digits is only possible on the basis of a proper understanding of the context.[13] An additional ambiguity is caused by the use of zodiacal signs beside degrees, often without any formal distinction whatever. For example [14]

$\overline{\varsigma\ \kappa\mu\kappa\zeta}$ means $6^s\,20;40,27°$

as becomes evident only if one follows step by step the computation in which it occurs.

The tables indicate, of course, the significance of all numbers by means of the heading of all columns, e.g.

ζώδια, μοῖραι, λεπτὰ ᾱ, δεύτερα

for "signs, degrees, minutes, seconds." Occasionally one finds also primes used for the consecutive fractions e.g.[15]

$\beta\ \overline{\iota\delta}\ \iota\delta'\ \lambda\zeta'\ \iota\alpha''\ \mu\beta'''$ for $2^s\,14;14,37,11,42°$.

[5] *Cf.* the bibliography, p. 44.

[6] *Cf.* Gildemeister in Usener, Kl. Schr. III, p. 340 f. The author, who is quoted fol. 291r 13, is al-Khāzinī; *cf.* Kennedy, Survey, p. 129, No. 27 and p. 159 f.

[7] His "τοῦ αὐθημερινοῦ τῶν ἰχθύων," however, is due to a misinterpretation of the symbol for ἀστέρων as the sign for pisces.

[8] For a general description and a table of contents *cf.* appendix 17 (p. 31).

[9] All folio references in the following concern Cod. Vat. gr. 1058 if not stated otherwise.

[10] *Cf.* Kennedy, Survey.

[11] fol. 316 to 320.

[12] fol. 293r,12.

[13] The number given above is obviously the length of the lunar year.

[14] fol. 261r,19.

[15] fol. 335r.

Beside the Greek alphabetic numerals, there also appear Hindu numerals [16] in forms depicted in table 2, but used only for year numbers, never in computations.

TABLE 2

1	ı	6	ϥ
2	ᴦ	7	V
3	ᴦᴜ or μ	8	Λ
4	ᒍ or ᒍ	9	9
5	ϑ or ϑ	0	δ or δ or ϥ

Obviously these signs were not too familiar to the scribe of our manuscript since one often finds on the margin or between the lines the equivalent in Greek numerals.[17]

A symbol for zero occurs, of course, both for Greek and for Hindu numerals.[18] The Greek symbol is either the customary ο or a sign which looks like ɥ, practically indistinguishable from the Hindu numeral for 6 (*cf.* table 2). Again it is only the context which furnishes the proper reading, though one can be sure that in computations with Greek numerals only the reading zero is possible.[19] Repeated zeros are written whenever required, e.g.[20]

$\overline{\text{ɥɥ } \kappa\epsilon\ \nu\delta}$ for $0^s\,0;25,54°$

or [21]

$\overline{\gamma\ \text{ɥɥɥ}}$ for $3^s\,0;0,0°$.

The form ɥ undoubtedly comes from Arabic forms for zero,[22] which, in turn, go back to Greek forms known to us from papyri. The Hindu form is simply ο without a bar.[23] In the writing of year numbers there often occur hybrid forms like υνο (for 450) or χλō (for 630).[24]

[16] I avoid the misleading term "Arabic numerals," for Hindu numerals. When I mention "Arabic" numerals, I mean Arabic alphabetic numerals.

[17] fol. 392^v or fol. 323^v,8.

[18] Contrary to popular belief, the use of a zero symbol in Greek astronomy precedes by centuries the beginning of Hindu astronomy. Its real origin is found in Babylonian astronomy. *Cf.*, e.g., my *Exact sciences in antiquity*, 2nd ed., Providence, 1957, index p. 240.

[19] Heiberg, Byz. Anal., 163 read βɥ as 20 which is, of course, impossible, since 20 in Greek numerals is never written as 2 plus zero but only as κ. The correct reading is sexagesimal 2,0 (= 120).

[20] fol. 264^v,23.

[21] fol. 262^v,14.

[22] *Cf.* Rida A. K. Irani, Arabic numeral forms. *Centaurus* 4: 1-12, 1955; also Neugebauer, Ex. Sci., p. 14.

[23] fol. 371^v.

[24] fol. 333^r, column 1,1 and col. 2,6. For examples *cf.* my paper on al-Kaid, *Jour. Amer. Orient. Soc.* 77: 313, 1957.

GLOSSARY OF TECHNICAL TERMS

See also the *Index verborum* (p. 42) and the list of *Arabic and Persian Technical Terms* (p. 41)

αἰλάτζ

starter, corresponding to the ἀφέτης of Hellenistic astrology. The origin of the term αἰλάτζ seems to be unknown (*cf.* Bīrūnī, Astrol. p. 323 n. 2). Our text says τὸ αἰλὰτζ κατ' ἰνδούς.[1] Persian *haylāj*, Latin *hyleg*; *cf.* Nallino, Batt. II, p. 355.

[1] fol. 312r,6.

ἄκρον

pole

ἀνάβασις τοῦ τόπου τῶν ἄκρων τῆς σφαίρας τῶν ζωδίων ἤγουν τῶν ἄκρων τῆς κερκίδος δι' ἧς κινεῖται ἡ σφαῖρα[1] "altitude of the place of the poles of the sphere of the zodiac or of the poles of the axis around which the sphere (of the zodiac) rotates."

ἀνάβασις τοῦ τόπου τῶν ἄκρων τῆς κερκίδος[2] altitude of the pole (of the ecliptic).

[1] fol. 299r,5 f.
[2] fol. 299r,10 f.

ἀλληλουχία

column of numbers in a table, e.g. εἰς τὴν ϵ' ἀλληλουχίαν[1] "in column 5."

[1] fol. 329r,13.

ἀνά

ἀνὰ $\overline{αα}$ ἡμέραν[1] day by day.

ἀνὰ $\overline{κκ}$[2] in steps of 20 (years).

ἀφαιροῦνται ἀνὰ $\overline{σι}$ $\overline{σι}$[3] reduce modulo 210.

[1] fol. 337v/338r; table for daily motions.
[2] fol. 272v,9 ff.: κατ' ἐναντίον τῶν χρόνων τῶν ῥωμαίων τῶν ἀνὰ $\overline{κκ}$ γίνεται εἰσέλευσις εἰς τὸ κανόνιον τῶν εἰκοσαετηρίδων.
[3] fol. 278v,6.

ἀνάβασις

ἀνάβασις βορεία (resp. νοτία)[1] *increasing and to the north* (resp. *to the south*) of the equator; in table of declinations.

ἥμισυ τῆς ἀναβάσεως τῆς σφαίρας[2] *ascending half of the sphere*, i.e., semicircle of the ecliptic between upper and lower culmination, containing the ascendant.

κύκλος τῆς ἀναβάσεως[3] *altitude circle.*

ἀνάβασις τοῦ ☉ κατὰ τὸ μέσον τῆς ἡμέρας[4] or ἀνάβασις ☉[5] or ἐσχάτη ἀνάβασις τοῦ ☉[6] *noon altitude of the sun.*

ἐσχάτη ἀνάβασις εἰς τὸν κύκλον τοῦ μέσου τῆς ἡμέρας[7] *meridian altitude* (of a star).

τετελειωμένη ἀνάβασις[8] or τελεία ἀνάβασις[9] *complement of altitude*, i.e., zenith distance.

ἀνάβασις τῆς ἡμέρας[10] *length of daylight*, expressed in degrees or in hours.

σημεῖον τῆς μοίρας τῆς ἀναβάσεως *see* appendix 14.

ἀνάβασις πρώτη = 60, ἀνάβασις δευτέρα = 60^2 for the order of sexagesimal numbers; e.g. in column headings: ἀνάβασις β / ἀνάβασις ᾱ / ἡμέραι / λεπτά.[11] Corresponding Arabic terminology: "first, second, . . . elevated."[12]

see also μορφοῦ.

[1] fol. 430v.
[2] fol. 311v,16 f.; 314r,12.
[3] fol. 360r: πλεῖον καὶ ἔλαττον ἀπὸ τῆς ὄψεως εἰς τὸν κύκλον τῆς ἀναβάσεως *parallax in the altitude circle.*
[4] fol. 265r,9.
[5] fol. 431v.
[6] fol. 288r,5.
[7] fol. 288r,20 f.
[8] fol. 299v,18.
[9] fol. 360r; heading of first column.
[10] fol. 265v,7,10.
[11] fol. 371v, 372r.
[12] *Cf.*, e.g., Luckey, Rechenkunst p. 41.

ἀναφοραί

see τόπος τῆς τύχης and appendix 4

ἀνώμαλος

πλέον καὶ ἔλαττον ἤτοι ἀνώμαλον[1] *increment* of equation of Mercury.

λεπτὰ ἀνώμαλα[2] or πλέον ἔλαττον ἰδίου[3] *parallax as function of anomaly*; *cf.* appendix 10.

[1] fol. 418r.
[2] fol. 428v col. 8.
[3] fol. 358v col. 12.

ἁπλός

single

δάκτυλα ἁπλᾶ *see* δάκτυλος

ἀποκατάστασις

for lunar eclipses:

ὥρα τῆς ἀποκαταστάσεως[1] end of the total phase.

τελεία ὥρα τῆς ἀποκαταστάσεως[2] end of the partial phase.

for solar eclipses:

ἡ τελεία ὥρα τῆς ἀποκαταστάσεως τῆς ἐκλείψεως[3] end of the partial phase.

[1] fol. 269r,27 f.
[2] fol. 269r,31.
[3] fol. 270v,27.

ἀτελής

ἀτελὲς ὕψωμα τοῦ ἡλίου motion of the solar apogee since

epoch, in the same time also motion of all planetary apogees; equal to motion of precession. *Cf.* appendix 3 and note 4 there.

δάκτυλος ἀτελής *linear digit* of eclipse magnitudes; *cf.* δάκτυλος.

αὐθημερινόν [1]

true longitude of the sun, the moon, or of a planet.

synonymous with ἐποχή: περὶ τῆς εὑρέσεως τῆς τοῦ ♂ ἐποχῆς — τὸ ἐξελθόν ἐστι τὸ αὐθημερινὸν τῆς ζητουμένης ἡμέρας τοῦ ♂ [2] synonymous with τόπος: ὁ τόπος τῆς ☾ ἤγουν τὸ αὐθημερινόν.[3]

λεπτὰ τοῦ αὐθημερινοῦ [4] a coefficient of correction for anomaly in lunar eclipses; *cf.* appendix 15.

1 Misread by Tannery as ἀνθημερινόν (Mém. sci. II, p. 318, as well as in original publication).
2 fol. 261^r,1 and 12 respectively.
3 fol. 300^r,14.
4 fol. 303^r,29; 303^v,2 f.; 358^v, col. 11.

βαθμός

magnitude of fixed stars: οἱ μείζονες ἀστέρες μέχρι καὶ τῶν μικροτέρων ϛʹ βαθμοὶ ἐτέθησαν.[1]

one sexagesimal order, i.e. 60 or 1/60:

παρ' ἕνα βαθμὸν πλέον κρατεῖται ἤγουν ἄνω [2] to raise a result by one sexagesimal place (i.e., multiply by 60); also παρ' ἕνα βαθμὸν κρατεῖται.[3]

παρ' ἕνα βαθμὸν ἔλαττον κρατεῖται [4] to lower a result by one sexagesimal place (i.e., divide by 60); [5] also κρατεῖται ἔλαττον ἑνὸς βαθμοῦ [6] or ἐκρατήθησαν παρ' ἕνα βαθμὸν κάτω.[7]

1 fol. 317^r,28 f.
2 fol. 290^v,27 f.
3 fol. 285^r,13.
4 fol. 290^v,16,23.
5 also incorrectly used when multiplication by 60 is required, e.g., fol. 285^r,25 f.
6 fol. 287^r,2 f.
7 fol. 285^v,11 f.

βάθος *see* πλάγιον

γενικός

γενικὰ λεπτά [1] *proportional parts,* expressed in sexagesimal fractions, i.e., coefficients of interpolation; in particular the coefficients which modify the epicyclic equation for the moon and the planets at intermediate distances; [2] *cf.* appendix 2 and 3.

λεπτὰ γενικὰ πρῶτα and δεύτερα [3] first and second type of coefficients of interpolation in lunar and planetary tables; *cf.* appendix 6 and *s.v.* μερισμός.

1 fol. 340^v to 358^r, column δ.
2 Almagest V,8: διαφορὰ ἑξηκοστῶν (Heiberg I, p. 390) and XI,11: ἑξηκοστὰ ἀφαιρέσεως (Heib. II, p. 436) respectively.
3 fol. 389^r.

γραμμή

see εὐθεία γραμμή.

see πλάτος μετὰ τῆς ὀρθῆς γραμμῆς.

γωνία

γωνία τοῦ μήκους [1] *angle* between the circle of altitude and the circle of latitude (for *longitudinal* parallax).

γωνία τοῦ πλάτους [1] *angle* between the circle of altitude and the ecliptic (for *latitudinal* parallax).

1 fol. 299^v,28 f.

δάκτυλος

digit of eclipse magnitudes.

δάκτυλοι or δάκτυλα τῆς διαμέτρου [1] or δάκτυλα ἁπλᾶ [2] twelfths of diameter (of sun or moon).

δάκτυλοι τῆς ἐπιφανείας [3] or δάκτυλα τέλεια [4] twelfths of area (of sun or moon); Almagest VI,8 [5]: δάκτυλοι ἡλίου and σελήνης respectively.

δάκτυλος ἀτελής *linear digit* of eclipse magnitudes; contrast: δάκτυλος τέλειος *area digit.*[6]

see also κρύψις δακτύλων *and* ὄρθωσις δακτύλων.

1 fol. 359^v and 428^v respectively (for lunar eclipses).
2 fol. 428^v (for solar eclipses).
3 fol. 359^v.
4 fol. 428^v.
5 Heib. I, p. 522.
6 Cod. Laur. Plut. 28, cod. 14 (unpublished) fol. 33^v, 9 f.: ἐν τῷ περσικῷ δακτύλων τελείων μὲν ῑ ἀτελῶν δὲ ῑ ιʹ. Similarly fol. 33^r, 23 and 33^v, 26. All three cases concern solar eclipses, according to Persian Tables (A.D. 1379 May 16; 1376 July 17; 1386 Jan. 1 respectively). The author of the Greek treatise (fol. 18 to 33) is perhaps Isaac Argyros, because the epoch of A.M. 6876 (= A.D. 1367) Sept. 1, used fol. 18, is the same as in the Easter Tables of Isaac Argyros. *Cf.* Halma, Table pascale du moine Isaac Argyre, Paris 1825, p. 1 ff., p. 12 ff. and Heiberg, Byz. Anal. p. 171 (cod. Vat. gr. 1058, fol. 246 to 249).

διὰ c. gen. λαμβάνειν

to take as function of.

αὕτη διὰ τῆς μέσου ὁδοῦ λαμβανομένη καὶ προστιθεμένη αὐτῇ ἀφαιρεῖται τῆς ἰδίας [1] "this is taken as function of the mean longitude and added to it, but subtracted from the anomaly."

1 fol. 394^v, marg.; for other examples *cf.* διακρίνω.

διακρίνω

to modify, to adjust.

αὕτη διὰ τῆς μέσου παρόδου λαμβανομένη καὶ αὐτῇ μὲν προστιθεμένη τῆς δὲ ἰδίας ἀφαιρουμένη ἑκατέραν διακρίνει [1] "this is to be taken as function of the mean longitude; added to it (the mean longitude) and subtracted from the anomaly respectively, makes each (longitude and anomaly) adjusted."

διὰ τῆς διακεκριμένου ἰδίας [2] "as function of the adjusted anomaly."

μοῖραι ἰδίου διακεκριμένου [3] "degrees of the adjusted anomaly."

ἡ μέση διακεκριμένη πάροδος [4] "the adjusted mean longitude" ἡ μέση διακριθεῖσα [5] "the adjusted mean (longitude)."

interchangeable with τέλειος: ἡ τελεία μέση ἤτοι διακεκριμένη κίνησις.[6]

[1] fol. 416v, marg.
[2] fol. 418r, marg.
[3] fol. 388v.
[4] fol. 388v, marg.
[5] fol. 409v, marg., 417v, marg.
[6] fol. 262v, marg. 2; *cf. also* fol. 412v: μοῖραι μέσου τελείας ὁδοῦ contrasted with fol. 413r: μοῖραι μέσου ὁδοῦ διακεκριμένης.

διάστασις

διάστασις τῶν ἀστέρων ἀπὸ τοῦ κύκλου τοῦ κατὰ τὸν νυχθήμερον κινουμένου [1] *distance* of stars from the equator; for the definition of "distance" *cf.* appendix 13.

[1] fol. 286v,10 f.

ἐγγύς, ἔγγιστα

approximately; particularly for rounded numbers—*passim.*

μῆκος ἐγγύτερον *see* μῆκος.

εἰσέλευσις εἰς τὰ κανόνια

entry in tables—*passim.*

ἐκβολή

determination of any quantity, e.g. περὶ τῆς ἐκβολῆς τοῦ αὐθημερινοῦ τοῦ ☉ [1] "on the determination of the true longitude of the sun"; parallel: περὶ τῆς εὑρέσεως τῆς τοῦ ☉ ἐποχῆς.[2]

[1] fol. 261r,13 *et passim.*
[2] fol. 261r,1.

ἐκτλῆφι

increment: τὸ ἐκτλῆφι ἤτοι τὸ πλεῖον καὶ ἔλλατον [1] from Arabic *ikhtilāf*; *cf.* Nallino, Batt. II, p. 330.

τὸ ἐκτλῆφι μανδὰρ ἤτοι τὸ πλέον καὶ ἔλαττον τῆς ὄψεως [2] *parallax*; from *ikhtilāf manẓar*; *cf.* μανδάρ.

κανόνια τοῦ ἐκτλῆφι μαδὰλ ἤτοι τῶν παραλλάξεων τοῦ πλείονος καὶ ἐλάττονος τῆς ὄψεως [3] *adjusted parallax*, i.e., difference between lunar and solar parallax; *cf.* μαδάλ.

[1] fol. 262v,29.
[2] fol. 300r,1 f.
[3] fol. 269v,20 f.

ἔλλαμψις *see* φῶς

ἐναλλαγή

κανόνιον τῶν ὡρῶν τοῦ κινήματος τῆς μέσης τῆς ἐναλλαγῆς [1] is a table which gives all multiples, from 1 to 60, of 0;24,20,57,38°. This table is followed by a table for the περίσσεια τῆς περιφορᾶς τοῦ χρόνου τοῦ ☉ [2] i.e. the "excess of revolution" (*cf.* appendix 16) and then by a table τῆς μέσης ὁδοῦ τῆς ἐναλλαγῆς.[3] The latter table can be described as follows: we use as independent variable true solar positions, in steps of 1°, counted from the perigee P (λ_p in fig. 1); we tabulate against

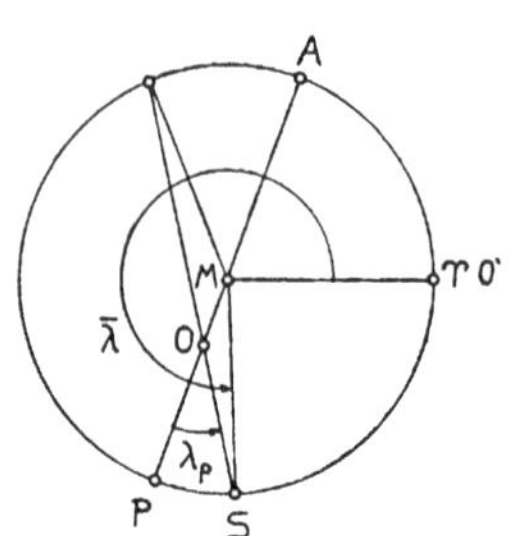

FIG. 1

this variable the corresponding mean solar longitudes $\bar{\lambda}$ counted from ♈ 0°. Then the increment of $\bar{\lambda}$ has a maximum of 1;2,5° at the apogee (about 2s 27°) and a minimum of 0;57,55° at the perigee (about 8s 27°). In short we have here a kind of "*inverse*" table of anomaly and this reciprocity is probably expressed in the title.

The parameter of the first table satisfies the relation

$$24;20,57,38 \cdot 15 = 6,5;14,24,30.$$

If the units on the right-hand side were days, one could interpret it as a value for the length of the tropical year. The text [1] gives as units of the left-hand side degrees; the independent variable is denoted as περίσσεια "excess," probably counted from 0;1 to 1. This term, as well as the fact that the subsequent table gives the "excess of revolution," suggests that also these tables are related to astrological concepts.

[1] fol. 429r left.
[2] fol. 429r right.
[3] fol. 429v, 430r.

ἐνῶ

to *add*—*passim.*

ἕνωσις *addition.*[1]

[1] fol. 295r,3: ἀπὸ τῆς ἑνώσεως ἢ ἀφαιρέσεως.

ἐπουσία

increase since epoch: ἐνιαύσιαι (resp. μηνιαῖαι) ἐπουσίαι συνοδοπανσεληνιακαί [1] "yearly (monthly) increments for conjunctions and oppositions."

[1] Cod. Laur. Plut. 28 cod. 14 fol. 22r; *cf.* Almag. VI, 2 p. 465,11 Heib. and Pappus, Comm. p. 178,9.

ἐποχή

longitude—*passim*; *cf.* αὐθημερινόν.

ἰσσάς

from Arabic *ḥiṣaṣ proportional part, coefficient*; *cf.* Nallino, Batt. II p. 328. *See also* τακκαέκ.

ἰσσὰ σαὰτ μπότ: fol. 423ᵛ omits the Greek equivalent, but Vat. gr. 185 fol. 21ʳ gives μερισμοὶ τοῦ μήκους τῶν ὡρῶν *coefficients for the length of hours*. The numbers σ tabulated on fol. 423ᵛ are obtained by dividing 24^h by the length of daylight c which is expressed in equinoctial hours. The extrema are $\sigma = 2;24$ for $c = 10^h$ and $\sigma = 1;42,51$ for $c = 14^h$. The limits 10^h and 14^h are the classical values for shortest and longest daylight in Alexandria.

ἔσχατος

ἐσχάτη ἀνάβασις τοῦ ☉ [1] *noon altitude of the sun.*

ἐσχάτη ἀνάβασις εἰς τὸν κύκλον τοῦ μέσου τῆς ἡμέρας [2] *meridian altitude* (of a star).

[1] fol. 288ʳ,5.
[2] fol. 288ʳ,20 f.

εὐθεία γραμμή

used for *sphaera recta.*

τόπος τῆς τύχης τῶν ζωδίων μετὰ τῆς εὐθείας γραμμῆς [1] *right ascension*; *cf.* τόπος.

[1] fol. 284ᵛ,12,16; 286ʳ,15 *et passim.*

ζῆζι

astronomical work, equivalent of σύνταξις [1] from Arabic *zīj.*

[1] fol. 261ᵛ,13: τὸ ζῆζι τόδε ἤτοι ἡ σύνταξις.

ζουμπρᾶ

the lunar mansion *zubra* (θ and δ Leonis).[1]

[1] fol. 333ᵛ, left, 8. *Cf.* Nallino, Scritti V, p. 178; Batt. II, p. 336.

ζώνη

ἡ τελεία τῆς ἡμέρας ζώνη ἤγουν ὁ κατὰ τὸ νυχθήμερον κινούμενος κύκλος [1] the celestial *equator.*

ζώνη τῆς σφαίρας [2] *equator of a sphere.*

[1] fol. 284ʳ,3 f.
[2] fol. 318ᵛ,1 ff.

ἥμισυ

ἥμισυ τῆς ἀναβάσεως (καταβάσεως) τῆς σφαίρας *see* ἀνάβασις (κατάβασις).

θαβὰν ὁ ἀλεξανδρηνός or θαβανῆ

Theon of Alexandria, cited for tables of parallax.[1]

[1] fol. 300ʳ,5 f. and 360ᵛ, 361ʳ: ἔργον τοῦ θαβανῆ.

θεμέλιον

τὰ θεμέλια τῶν ἀστέρων εἰς τὴν ἀρχὴν τοῦ ἔτους ($\overline{\text{φμα}}$) τῶν περσῶν [1] *values at epoch,* corresponding to *radix* in Western tables; *cf.* appendix 1.

τόπος τοῦ θεμελίου τοῦ ἀστέρος [2] *geocentric position* of a celestial body, excluding parallax (contrast: τόπος τῆς ὄψεως). In computations an intermediary result is often called a θεμέλιον i.e., a number on which further computations are based; e.g., if δ is the distance from the boundary of a zodiacal sign of a planet at the nearest noontime before or after crossing it, then $\delta/24$ is called the θεμέλιον.[3]

[1] fol. 376ʳ, 377ʳ: κανόνιον τῶν θεμελίων etc.
[2] fol. 320ʳ,10; also Vat. gr. 211, fol. 120ʳ, upper figure.
[3] fol. 266ᵛ,4 f.

θεωρία *see* τόξον τῆς θεωρίας

ἴδιος

ἰδία κίνησις [1] or simply ἴδιον [2] *motion in anomaly* of the moon or the planets; equivalent: χασσᾶ (or χατζᾶ [3]) μαντὰλ ἤτοι ἰδία κίνησις; [4] *cf.* χασσᾶ.

ἴδιον τέλειον [5] *anomaly,* counted from apparent apogee G of the epicycle; *cf.* appendix 2, fig. 4 and appendix 3, fig. 5.

πλέον ἔλαττον ἰδίου [6] *parallax* as function of anomaly; *cf.* appendix 10. Equivalent: λεπτὰ ἀνώμαλα.[7]

[1] fol. 337ᵛ, 338ʳ.
[2] fol. 375ᵛ.
[3] fol. 264ʳ,26.
[4] fol. 263ʳ,28.
[5] fol. 325ʳ,12 f.
[6] fol. 358ᵛ, col. 12.
[7] fol. 428ᵛ, col. 8.

ἰντεέ

from Arabic *intihā' end*; *cf.* Bīrūnī, Astrol. § 517 (Wright, p. 320).[1]

[1] fol. 315ᵛ,12: ἐκεῖνο τὸ ζώδιον ἰντεὲ καλεῖται.

καιρός

time, duration.

κανόνιον τοῦ καιροῦ τῆς ἐκλείψεως τῆς ☾ [1] table for the time from first contact and from beginning of totality to the middle of the eclipse.

ὥρα τοῦ καιροῦ [2] half duration of totality.

time degrees.

νυχθήμερον $=$ 360 καιροί [3] $=$ 24 hours.

passim for the *moment* under consideration.

[1] fol. 427ʳ; for solar eclipses: fol. 428ᵛ.
[2] fol. 269ʳ,18,20.
[3] fol. 285ᵛ,24 f.

κάϊτ

from Sanscrit *ketu* and Persian *kaid*; a fictitious comet in retrograde motion with a period of 144 years.[1]

[1] fol. 365ʳ, 374ʳ. *Cf.* W. Hartner, *Les Conférences du Palais de la Découverte,* sér. D, No. 36, 1955; E. S. Kennedy, *Jour. Near Eastern Studies* **16**: 44-51, 1957; O. Neugebauer, *Jour. Amer. Orient. Soc.* **77**: 211-215, 1957.

καπισά

intercalary years;[1] from Arabic *kabīsa*; *cf*., e.g., Bīrūnī, Astrol. § 271.

[1] *passim*, e.g. fol. 332r, left lower table: total of days = 366. *Cf.* also Bullialdus, Astr. Philol., p. 213: Καπισα ἤτοι δίσεκτον, and *CCAG* 1, p. 88,21 f.

καταβασις

ὁ καιρὸς τῆς καταβάσεως τοῦ ☉[1] *the moment of sunset*.

ἥμισυ τῆς καταβάσεως τῆς σφαίρας[2] *descending half of the sphere*, i.e. semicircle of the ecliptic from lower to upper culmination, containing the setting point.

κατάβασις βορεία[3] (resp. νοτία) *decreasing and north* (resp. *south*) of the equator.

[1] fol. 307r,1; table fol. 363r to be corrected by Vat. gr. 211 fol. 153r.
[2] fol. 314r,29.
[3] fol. 430v.

καταβατόν

column, in particular for the column of 29 to 32 days in an ephemeris.[1]

[1] fol. 328r,4,8 f.: εἰς τὸν τελευταῖον ψῆφον τοῦ καταβατοῦ. Otherwise the most common expression for column is σελίδιον.

κέντρον

for the moon:

elongation as well as *double elongation* from the sun (*cf*. appendix 2, 5, and 6). *See also* μετάβασις.

for the planets (*cf*. appendix 3 and figure 5):

κέντρον: $\bar{\kappa} = \bar{\lambda} - \lambda_A$ i.e. distance, seen from the equant E, of the center C of the epicycle from the apogee A.

κέντρον τέλειον: $\kappa = \bar{\kappa} - \omega_1$ (ω_1 = ὄρθωσις πρώτη) the distance of C from A, seen from O.

also used like *column* (of numbers), e.g., κατεναντίον οὖν τῶν η κβ εἰς τὸ βον κέντρον εὑρέθη ψῆφος ς κθ[1] "opposite to 8s 22° in the second column, one finds (in the third column) the number 6;29." Actually this is only an abbreviated terminology "in the second column which is headed κέντρον."

[1] fol. 329v,10,11 *et passim* in similar context.

κερκίς

axis; *see* ἄρκον.

κεφάλαιον

τὰ ε̄ κεφάλαια[1] *the five sections* from which a mean position of one of the celestial bodies is computed, namely: (1) groups of years (e.g. 25 Egyptian years, or 30 Arabic years, etc.) (2) single years (3) months (4) single days (5) hours.

[1] *passim*, e.g. Theon, Handy Tables, Halma I, p. 45,5 or p. 47; Usener, *Kl. Schr.* III, p. 302,28 *et passim*. Heiberg, Ptolem. Opera II, p. CXCIII f. 58r incorrectly rendered as τῶν ε̄ κλιμάτων; same error Cod. Vatic. Gr. I. p. 255.

κίνημα τῶν ὡρῶν

reciprocal value of the hourly increase of anomaly of a planet.[1]

[1] fol. 423r, e.g. for Saturn 0;25,12,14,40h/°, since the hourly increase of the anomaly amounts to 0;2,22,49 . . .°

κίνησις

mean motion, in contrast to ἴδιον—*passim*.

κόμπος[1]

the *lunar nodes*: ἡ ☾ πλησίον ἐστὶ τῶν κόμπων ἤγουν τοῦ ἀναβιβάζοντος ἢ τοῦ καταβιβάζοντος.[2]

ὅτι μέσον τῶν κόμπων καὶ τῶν μοιρῶν τῆς ☾ ἔλαττον ὀφείλει εἶναι τῶν ιβ μοιρῶν[3] condition for lunar eclipse: that the distance between the nodes and the place of the moon be less than 12°.

ἡ σφαῖρα τῶν κόμπων τῆς ☾[4] the *eclipse* (contrast: σφαίρα μετακλίσεως[4] for the inclined lunar orbit).

[1] *Sic*, for κόμβος *knot*. Spelling certain.
[2] fol. 320v,31–321r,1.
[3] fol. 268r,10 f.
[4] Vat. gr. 211, fol. 120v, lower figure.

κουσούφ

solar eclipse: κουσοὺφ λέγεται κατὰ πέρσας ἤτοι ἔκλειψις· τῆς δὲ ☾ χουσοὺφ ἤτοι κρύψις[1] from Arabic *kusūf* and *khusūf* respectively [Kennedy]; *cf*. Nallino, Batt. II, p. 352; Bīrūnī, Astrol. § 255 ff.

[1] fol. 269v,9 f.

κρύψις

lunar eclipse; *see* χουσούφ.

κρύψις δακτύλων[1] *eclipse magnitude*, measured in linear digits.[2]

[1] fol. 359v, tabulated as linear function of the latitude.
[2] i.e. twelfths of the diameter (not of the area).

κυκλικὴ περίοδος

under this heading fol. 375r gives a list of numbers which are obtainable from the corresponding list of daily motions by multiplication with a factor 6 (probably not to be interpreted as 6,0 since one then would expect reductions mod. 6,0). I do not know the reason for this terminology nor can I explain the purpose of the corresponding table.

κύκλος

ὁ μικρὸς κύκλος[1] the *epicycle*.

κανόνια τοῦ πλείονος καὶ ἐλάττονος ἀπὸ τῆς ὄψεως εἰς τὸν κύκλον τῆς ἀναβάσεως[2] table of parallax in the *circle of altitude*.

κύκλος τῆς ὀρθώσεως τῆς ἡμέρας[3] or κύκλος τῆς ὀρθώσεως[4] or κύκλος τῆς ἡμέρας[5] the celestial *equator*.

τέλειος κύκλος τῆς ἡμέρας [6] or ἡ τελεία τῆς ἡμέρας ζώνη ἤγουν ὁ κατὰ τὸ νυχθήμερον κινούμενος κύκλος [7] the celestial *equator*.

κύκλος τοῦ μέσου τῆς ἡμέρας [8] *meridian*.

κύκλος τῆς κινήσεως *see* πλάτος τοῦ κύκλου τῆς κινήσεως.

[1] fol. 301ʳ,4; 368ᵛ, col. γ.
[2] fol. 360ʳ.
[3] fol. 286ᵛ,26.
[4] fol. 286ᵛ,14.
[5] fol. 286ᵛ,24.
[6] fol. 285ʳ,3.
[7] fol. 284ʳ,3 f.
[8] fol. 284ʳ,15 f.; 286ʳ,22.

λεπτά

minutes, or higher sexagesimal fractions; in particular λεπτὰ δεύτερα.

seconds—passim. Also used in the singular, e.g., εὑρέθη $\overline{ια}$ $\overline{κη}$ $\overline{ν}$ $\overline{θ}$ $\overline{κγ}$ ζωδ. μοῖ. λεπτὸν [1] $= 11^s\,28;50,9,23^\circ$.

λεπτὰ ἀνώμαλα *see* ἀνώμαλος.

λεπτὰ γενικὰ πρῶτα and δεύτερα *see* γενικὰ λεπτά and μερισμός.

λεπτὰ τοῦ αὐθημερινοῦ [2] a coefficient of correction for anomaly in lunar eclipses; *cf.* appendix 15.

λεπτὰ ἀνώμαλα [3] or πλέον ἔλαττον ἰδίου [4] *parallax as function of anomaly*; *cf.* appendix 10.

λεπτὰ σφαίρας *see* σφαῖρα.

[1] fol. 261ʳ,22; similar 23.
[2] fol. 303ʳ,29; 303ᵛ,2 f.; 358ᵛ, col. 11.
[3] fol. 428ᵛ, col. 8.
[4] fol. 358ᵛ, col. 12.

λογίζομαι

for rounding of numbers, e.g.: καὶ ἐγένοντο $\overline{δ\ νθ}$ ἤτοι ε′ τῶν νθ ἐν λογισθέντων [1] i.e., $4;59 \approx 5$. I owe this reading to a discussion with Dr. G. Stamires.

[1] fol. 329ʳ,20. *Cf.* fol. 263ᵛ, 10: τῶν λς′ ὡς α′ λογισθέντων.

μαδάλ

equated; from Arabic *mu'addal* [Kennedy].

κανόνια τοῦ ἐκτλῆφι μαδὰλ ἤτοι τῶν παραλλάξεων τοῦ πλείονος καὶ ἐλάττονος τῆς ὄψεως [1] *adjusted parallax*, i.e. difference between lunar and solar parallax.

ἐκεῖνο ψῆφός ἐστι τοῦ μαδάλ [2] "this is the value of the adjusted parallax."

cf. also χασσᾶ μαντάλ.

[1] fol. 269ᵛ,20 f.
[2] fol. 360ʳ,29.

μακόμ

corrected; from Arabic *maqūm* [Kennedy]; *cf.* Nallino, Batt. II, p. 350.

πλάτος ἐστὶ τοῦ μακὸμ ἤτοι στερεόν [1] *final latitude*, i.e. corrected for parallax.

[1] fol. 269ᵛ,29.

μανδάρ

vision, from Arabic *manẓar*. Latin *elmandhar*; *cf.* Suter, Khw., Tab. 77 and p. 244.

τὸ ἐκτλῆφι μανδὰρ ἤτοι τὸ πλέον καὶ ἔλαττον τῆς ὄψεως [1] *parallax*; from *ikhtilāf manẓar*; *cf.* Bīrūnī, Astrol. § 263; Nallino, Batt. II, p. 330; Suter, Khw., p. 243 *s.v. ihtilef*.

[1] fol. 300ʳ,1 f.

ματαλέ

τὸ ματαλὲ τῆς διαμέτρου τοῦ ☉ [1] *rising time* of the point opposite the sun; from Arabic *maṭāla'*; *cf.* Nallino, Batt. II, p. 342.

[1] fol. 369ʳ,8.

μερίζω

divide see τηρέω.

μερισμός

coefficient of interpolation for intermediary positions of the lunar epicycle; *cf.* appendix 6 and fig. 9*c*. In particular

$$\text{λεπτὰ πρῶτα τοῦ μερισμοῦ}^{1} = c_1(\kappa)$$
$$\text{λεπτὰ δεύτερα τοῦ μερισμοῦ}^{1} = c_2(\kappa).$$

Note that here πρῶτα and δεύτερα do not concern different sexagesimal orders but case distinctions; *cf. also* γενικὰ λεπτά.

μερισμὸς τῶν λεπτῶν [2] coefficient of interpolation of planetary latitudes (Almagest XIII,5: ἑξηκοστά); *cf. also* τακκαέκ.

μερισμοὶ τοῦ μήκους τῶν ὡρῶν *see* ἐσσὰ σαὰτ μπότ.

[1] fol. 384ʳ.
[2] fol. 392ʳ, 398ᵛ.

μέρος

μέρη τῆς ἡμέρας *fractions of a day* of the excess of one tropical year over 365 days. In the Maliki calendar this excess is 53/220 of one day ($= 0;14,27,16,\ldots{}^d$). The table fol. 332ʳ gives only the multiples of 53 (mod. 220) for single years, for multiples of 10 years and of 100 years. *Cf. also* appendix 16.

μέσος

μέση κίνησις *mean motion*; also *mean longitude*, counted either from the vernal point (*cf.* appendix 1) or from the apogee (*cf.* appendix 5).

μέσος ψῆφος *difference*, e.g. ὁ μέσος τῶν $\overline{λγ}$ καὶ $\overline{μα}$ ψῆφος ὃς ἦν η′ [1] or μετὰ τοῦ μέσου τῶν β′ κανονίων ψήφου ἤγουν τῆς περισσείας.[2]

μῆκος μέσον τῆς ζητουμένης πόλεως καὶ τοῦ μακκᾶ [3] *distance* between the city in question and Mecca.

μέσον τοῦ ζωδιακοῦ κύκλου καὶ τῆς τελείας τῆς ἡμέρας ζώνης [4] *distance* between ecliptic and equator.

[1] fol. 263r,30; similar fol. 263v,8.
[2] fol. 269r,12.
[3] fol. 290v,24 f.
[4] fol. 284r,2 f.

μετάβασις

used in the ordinary sense of *motion*: οὔτε γὰρ πλέον τῶν ιε′ μοιρῶν καὶ γ′ λεπτῶν προχωρεῖ ἡ τοιαύτη μετάβασις, οὔτε ἔλαττον τῶν ια′ μοιρῶν καὶ λα λεπτῶν [1] "the progress (of the moon) is never greater than 15;3° and never less than 11;31° (per day)."

μετάβασις ☉ εἰς μίαν ἡμέραν [2] resp. εἰς μίαν ὥραν [2] *daily velocity* resp. *hourly velocity* of the sun.

μετάβασις τῆς ☾ [3] *daily velocity* of the moon.

μετάβασις τελεία [4] *elongation* of the moon from the sun, interchangeable with κέντρον (*cf.* appendix 5); also for *double elongation*.[5]

[1] fol. 329v,3-5.
[2] fol. 358v.
[3] fol. 362v.
[4] fol. 270r,11.
[5] fol. 327r,22; fol. 328v,29 *et passim*.

μετάκλισις

declination, in particular of the sun.

μετάκλισις πρώτη [1] *first declination* = declination δ in the modern sense (δ_1 in fig. 2).

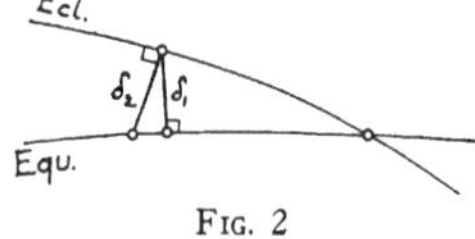

FIG. 2

μετάκλισις δεύτερα [1] *second declination* = arc of latitudinal circle between ecliptic and equator (δ_2 in fig. 2 and appendix 13).

σκιὰ τῆς πρώτης μετακλίσεως [2] = tan δ.

σφαῖρα τῆς μετακλίσεως [3] inclined (lunar) orbit.

μεγάλη μετάκλισις [4] *obliquity of the ecliptic*, also μετάκλισις ὅλη.[5]

[1] tabulated fol. 366r and 430v.
[2] fol. 431r.
[3] Vat. gr. 211 fol. 120v, lower figure.
[4] fol. 284r,2 ff.: ἡ μεγάλη μετάκλισις μέσον ἐστὶ τοῦ ζωδιακοῦ κύκλου καὶ τῆς τελείας τῆς ἡμέρας ζώνης ἤγουν τοῦ κατὰ τὸ νυχθήμερον κινουμένου κύκλου.
[5] fol. 301r,24.

μῆκος

distance, in particular the distance of a star from the equator, used parallel to διάστασις τῶν ἀστέρων ἀπὸ τοῦ τελείου κύκλου τῆς ἡμέρας.[1] For the definition of this "distance" *cf.* appendix 13.

μήκιστον μῆκος [2] *greatest distance* of the moon from the earth, i.e. at syzygies and in the apogee of the epicycle.

interval of time: τὸ μῆκος πρὸ τοῦ μέσου τῆς ἡμέρας καὶ μετὰ τὸ μέσον τῆς ἡμέρας [3] the *time interval* before or after noon.

μῆκος τῆς ὥρας [4] or ὥρα τοῦ μήκους [5] *time difference* from nearest noon, measured in hours, to reach the exact moment of conjunction or opposition at eclipses; *cf.* appendix 9.

ὥρα τοῦ μήκους [6] *time difference* between noon and the entry of the moon, or of the sun, or a planet, into a zodiacal sign.

ὥρα τοῦ μήκους τῆς ἡμέρας and ὧραι μήκους τῆς ἡμέρας [7] *longest daylight.*

μῆκος μέσου τοῦ ☉ καὶ τῆς ☾ [8] *difference of oblique ascension* of sun and moon; *cf.* appendix 4.

longitude of a celestial body, i.e. geocentric longitude measured in the ecliptic, eastward from the vernal equinox.[9]

geographical longitude, counted ἀπὸ τῆς ἄκρας θαλάσσης τῆς δύσεως [10] "from the outermost western sea." The Sanjarī tables are said [11] to be computed for a longitude of 90° but no specific city is mentioned. The version of these tables by Shams ad-Dīn of Buchara requires a transformation to 87° east, i.e., to the meridian of Buchara.[12] The tables of the σύνταξις ἀλαῆ (*zīj al-'Alā'ī*) are based on the meridian of 84° east.[13] A longitude of 82° is given to Daras.[14] The "Persian Tables" of Bullialdus and Usener are computed for the meridian of Τιβήνη 72° east.[15] Constantinople is placed at 49;50° east.[16]

for the moon:

μῆκος [17] *excess* of the second lunar equation over the simple equation for extreme distances.

μῆκος ἐγγύτερον [18] *excess* of the second lunar equation over the simple equation for intermediate positions of the epicycle; *cf.* appendix 2.

μῆκος ἤτοι κέντρον [19] *elongation* of the mean moon from the mean sun.

for the planets:

μῆκος ἐπιμηκέστερον [20] *decrease* of the epicyclic equation due to the removal of the epicycle from mean distance to maximum distance; also called πόρρω μῆκος.[21]

μῆκος ὁλοεπιμηκέστερον [22] *increase* of the epicyclic equation due to the removal of the epicycle from mean distance to minimum distance.

[1] fol. 285r,2 ff.; similar fol. 284r,7 f.
[2] fol. 360v to 361v.
[3] fol. 303v,28 f.
[4] fol. 268v,27; 270r,13 f.
[5] fol. 300r,21 f.
[6] fol. 265v,25; 266v,8.
[7] fol. 360v, 361r, and 361v respectively.

[8] fol. 331^{r},20.
[9] *passim.*
[10] fol. 323^{v},30 f.
[11] fol. 324^{r},3.
[12] fol. 323^{v},30. According to the list of famous cities, fol. 338^{v} (col. III,21) Buchara has the longitude 86;50° and the latitude 39;0°.
[13] fol. 261^{r},14; fol. 375^{r}; 379^{v}; 384^{v}, etc.
[14] fol. 272^{v},22.
[15] Usener, Kl. Schr. III, p. 371; *cf. also* p. 342 s). Also Vat. gr. 1058, fol. 130^{r},5; 150^{r} ff.; CCAG 5,3 p. 146, 8 f.
[16] fol. 261^{v},16 f., 338^{v}, column I,9.
[17] Tabulated fol. 340^{v} to 343^{r}, column 6.
[18] fol. 325^{r},28. Almagest V,8: ἐπικύκλου διαφορά (Heiberg I, p. 390).
[19] fol. 382^{v}, 384^{r}, explanatory notes to the headings of the tables; also fol. 262^{r},9 etc.
[20] fol. 343^{v} to 358^{r}, column 5. Almagest XI,11: διαφορὰ ἀφαιρέσεως (Heiberg II, p. 436, col. 5).
[21] fol. 326^{r},3.
[22] fol. 343^{v} to 358^{r}, column 7. Almagest XI,11: διαφορὰ προσθέσεως (Heiberg II, p. 436, col. 7).

μίγμα *see* μουδρούπ

μίγνυμι

to *add*

λαμβάνεται ἡ α′ ὄρθωσις καὶ μίγνυται μὲν αὐτῇ τῇ μέσῃ ἀφαιρεῖται δὲ τοῦ ἰδίου [1] $\kappa = \bar{\kappa} + \omega_1$ and $a = \bar{a} - \omega_1$ in the notation of appendix 7.

[1] fol. 262^{r}, marg. line 1.

μικρός

ὁ μικρὸς κύκλος [1] the *epicycle.*

[1] fol. 301^{r},4; 368^{v}, col. 3.

μοῖρα πλάτους *see* πλάτος

μονή

μοναὶ τῆς ☾ [1] *lunar mansions.*

[1] table: fol. 333^{v}; text; 280^{r},16 ff.

μορφοῦ

ἀπὸ τῶν ἀναβάσεων ὃ λέγεται κατὰ πέρσας μορφοῦ [1] steps of 60 (Arabic) years in computing precession; from Arabic *marfūʿ elevated* (for powers of 60); *cf.* Nallino, Batt. II, p. 335 or Luckey, Rechenkunst p. 41.

[1] fol. 286^{v},6.

μουδρούπ

μουδροὺπ ἤτοι μίγμα [1] *product*; *cf.* appendix 3. From Arabic *maḍrūb* [Kennedy].

[1] fol. 326^{r},5 f.; also 262^{v},24.

μπότ *see* ἰσσὰ σαὰτ μπότ

μυζαστῆ [1]

from μεγίστη (σύνταξις), i.e. the Almagest. *Cf.* Usener, Kl. Schr. III, p. 339.

[1] fol. 363^{v}.

ναχιζάκ

κανόνιον ναχιζὰκ κατὰ πέρσας [1] a concordance of Persian and other months; the Persian original seems to be unknown.

[1] fol. 276^{v}.

ντατίλ

correction, Latin *equatio.* From Arabic *taʿdīl*; *cf.* Nallino, Batt. II, p. 344.

ντατὶλ σαὰτ ῥοὲτ ἤτοι τῆς ὀρθώσεως τῶν ὡρῶν ἀπὸ τῆς ὄψεως [1] correction of the time of true conjunction (counted from noon) for longitudinal parallax; *cf.* appendix 9. From *taʿdīl-i sāʿat-i rūʾya* [Kennedy].

[1] fol. 269^{v},16; 270^{v},5. Table: fol. 425^{r}.

νυχθήμερον

ἡ τελεία τῆς ἡμέρας ζώνη ἤγουν ὁ κατὰ τὸ νυχθήμερον κινούμενος κύκλος [1] the *equator.*

[1] fol. 284^{r},3 f.

ὁδός

ἡ μέση ὁδός [1] *mean longitude.*

[1] fol. 394^{v}: αὕτη διὰ τῆς μέσου ὁδοῦ λαμβανομένη "this is taken as function of the mean longitude"; *cf. also* πάροδος.

οἴκημα

τὰ ιβ′ οἰκήματα [1] *the 12 loci,* counted in the sense of increasing longitude, beginning with locus 1 = ascendant. Thus locus 7 = setting point, locus 10 = culminating point.

[1] fol. 289^{r},19.

ὀρθός

ὀρθὴ ὥρα [1] *equinoctial hour.*

μὴ ὀρθὴ ὥρα [2] *seasonal hour.*

[1] fol. 286^{r},4.
[2] fol. 286^{r},7.

ὄρθωμα

correction, interchangeable with ὄρθωσις.

ὑπὸ τὴν ψηφηφορίαν τοῦ στερεοῦ ὀρθώματος τοῦ ☉ [1] "under the number of the final correction for the sun."

[1] fol. 261^{v},5 f.

ὄρθωσις

correction, corresponding to *taʿdīl* or *equatio.*

for the sun:

equation of center; *cf.,* however, στερεὰ ὄρθωσις (*s.v.* στερεός).

for the moon:

ὄρθωσις πρώτη [1] correction for apogee of eccenter. Almag. V,8: ἐκκέντρου προσθαφαιρέσεις ἀπογείου (Heib. I, p. 390, col. 3).

ὄρθωσις δεύτερα [2] simple equation, for epicycle at maximum distance. Almag. IV,10: προσθαφαιρέσεις (Heib. I, p. 337, col. 3) or V,8: πλάτους καὶ μήκους προσθαφαιρέσεις ἐπικύκλου (Heib. I, p. 390, col. 4).

ὄρθωσις στέρεα [3] final equation; *cf.* appendix 2

for eclipses and parallax:

ὄρθωσις σκιάσματος [4] coefficient indicating the increase in the diameter of the earth's shadow when the sun moves from the apogee to the perigee; min. = 0, max. = 0;1,40.

ὄρθωσις δακτύλων [5] correction of eclipse magnitudes, and ὄρθωσις ὡρῶν [5] correction of duration, due to the motion of the moon from the apogee of the epicycle toward the perigee.

ὄρθωσις τοῦ πλείονος καὶ ἐλάττονος τῆς ☾ [6] increase of the lunar parallax for syzygies and for the perigee of the epicycle. Almag. V,18: σελήνης δευτέρου ὅρου διαφορά (Heib. I, p. 442, col. 4).

ὄρθωσις τῶν ὡρῶν ἀπὸ τῆς ὄψεως [7] correction for longitudinal parallax of the time of conjunction, counted from noon. *Cf.* ντατὶλ σαὰτ ῥοὲτ and appendix 9.

for the planets:

ὄρθωσις πρώτη [8] correction of mean distance of the center of the epicycle from the apogee, if seen from 0 instead of from the equant. This combines the columns 3 and 4 in Almag. XI,11 (Heib. II, p. 436 ff.).

ὄρθωσις δεύτερα [9] epicycle equation assuming mean distance of the epicycle; Almag. XI,11: ἀνωμαλίας προσθαφαιρέσεις (Heib. II, p. 436 ff.).

ὄρθωσις στερέα [10] final equation; *cf.* appendix 3.

ὄρθωσις πλάτους [11] third component in the computation of latitudes of Venus and Mercury; *cf.* appendix 8.

ὄρθωσις τῶν ἡμερῶν [12] *equation of time*, normed to be always subtractive: ἡ ὄρθωσις τῆς ἡμέρας τεμμάχιον (*sic*) ἐστὶ τῆς ὥρας· τοῦτο τὸ τεμμάχιον ἀεὶ ἀφαιρεῖται.[13]

ὄρθωσις τῶν ἡμερῶν [14] (or τῆς ἡμέρας [15]) *ascensional difference*, i.e., increment of half of the length of daylight (measured in degrees) over 90°. *Cf.* appendix 12.

κύκλος τῆς ὀρθώσεως [16] or κύκλος τῆς ὀρθώσεως τῆς ἡμέρας [17] *celestial equator.*

ὄρθωσις τῆς περισσείας τῶν χρόνων [18] *correction for the excess of the years* (corresponding to the Arabic *faḍl al-dawr excess of revolutions* [Kennedy]), i.e., excess, measured in degrees, of the tropical year over 365 days; *cf.* appendix 16.

ὄρθωσις τοῦ σημείου *see* appendix 14.

[1] tabulated fol. 340ᵛ to 342ᵛ, col. 3.
[2] tabulated fol. 340ᵛ to 342ᵛ, col. 5.
[3] fol. 325ʳ,29 f.
[4] fol. 358ᵛ, col. 6.
[5] fol. 359ᵛ, col. 3 and 7 respectively.
[6] fol. 360ʳ, col. 4.
[7] fol. 269ᵛ,16; 270ᵛ,5; 425ʳ, col. 3.
[8] tabulated fol. 343ᵛ to 358ʳ, col. 3.
[9] tabulated fol. 343ᵛ to 358ʳ, col. 6.
[10] fol. 326ʳ,9.
[11] fol. 422ᵛ.
[12] tabulated fol. 339ʳ.
[13] fol. 339ʳ,1.
[14] fol. 366ᵛ and 373ᵛ.
[15] fol. 285ʳ,27.
[16] fol. 286ᵛ,14.
[17] fol. 286ᵛ,26.
[18] fol. 367ʳ, left.

ὄψις

τόπος τῆς ὄψεως τοῦ ἀστέρος [1] apparent position of a celestial body, i.e. including parallax (contrast: τόπος τοῦ θεμελίου).

πλέον καὶ ἔλαττον ὄψεως *see* πλέον.

[1] fol. 320ᵛ,9.

παραμονή

1494 Persian years of 365 days each, i.e., the years of the era Yezdegerd, corresponding to 1493 solar years.[1] Thus it takes 1494 Persian years until the beginning of a Persian month coincides again with the entry of the sun into Aries. The month which is in this position is called παραμονή.[2] Since 1494/12 = 124½ it takes alternatingly 124 and 125 years for a Persian month to be the first month of the solar year.

[1] fol. 276ʳ,10.
[2] fol. 276ʳ,23 f.; tabulated fol. 276ᵛ: μῆνες τῆς παραμονῆς.

πάροδος

μέση πάροδος [1] *mean longitude.*

ἡ μέση διακεκριμένη πάροδος [2] the adjusted mean longitude; *cf.* appendix 7.

ἰδία πάροδος [3] *anomaly.*

[1] fol. 416ᵛ, marg.: αὕτη διὰ τῆς μέσου παρόδου λαμβανομένη "this is taken as function of the mean longitude."
[2] fol. 388ᵛ, marg.
[3] fol. 396ʳ, marg.

πασιτά

ordinary year; [1] from Arabic *basīṭa.*

[1] *passim*, e.g. fol. 332ʳ left lower table: total of days 365. *Cf.* Bullialdus, Astr. Phil. p. 213: πασιτα ἤτοι ὁ μὴ ὢν δίσεκτος; also CCAG 1, p. 88, 23.

περίσσεια

difference, in particular tabular difference for interpolation; [1] equivalent: μέσος ψῆφος.

περίσσεια καὶ ἔλλειψις γίνεται [2] there occur *increments* or decrements.

περίσσεια τῆς ἡμέρας [3] (= twice the ὄρθωσις τῆς ἡμέρας) difference between 12^h or 180° and the length of daylight for a given solar longitude.

περίσσεια τῆς περιφορᾶς τοῦ χρόνου τοῦ ♂ or περίσσεια τῶν

χρόνων *excess of revolution*, i.e., excess of the tropical year over 365 days, measured in degrees; *cf.* appendix 16.

[1] *passim*, e.g. fol. 339ᵛ ff.
[2] fol. 308ʳ,12.
[3] fol. 285ᵛ,3,31.

περιφορά

περίσσεια τῆς περιφορᾶς τοῦ χρόνου τοῦ ☌ *see* περίσσεια.

πίπτω *see* ὥρα

πλάγιον

column in a table of two entries, e.g., τὸ ζώδιον ζητεῖται ἄνω τοῦ κανονίου καὶ αἱ μοῖραι ἐκ πλαγίου[1] "the zodiacal sign is to be taken from the top of the table and the degrees from the (left) *column*."

Contrast: βάθος lit. *depth*; e.g. ἡ μία (περίσσεια) γινομένη ἀπὸ τοῦ κατὰ βάθος ψήφου καὶ ἡ ἑτέρα ἀπὸ τοῦ ἐκ πλαγίου[2] "one (difference) originates from the numbers in a *row*, the other (from the numbers) in a *column*."

[1] fol. 284ᵛ,20.
[2] fol. 273ʳ,7 f.

πλάτος

latitude, measured on an arc perpendicular to the ecliptic.

for the moon:

μοῖρα τοῦ πλάτους[1] *argument of latitude*, i.e. distance from the ascending node to the moon.

πλάτος στερεόν *cf.* appendix 9.

for the inferior planets:

πρῶτον πλάτος[2] *inclination*, corresponding to ἐγκλίσεις in Almag. XIII,5 (Heib. II, p. 584).

δεύτερον πλάτος[3] *obliquity*, corresponding to λοξώσεις in Almag. XIII,5 (Heib. II, p. 584).

πλάτος μετὰ τῆς ὀρθῆς γραμμῆς[4] *geographical latitude*, ϕ.

πλάτος τέλειον[5] *complement of latitude*, $90 - \phi$.

πλάτος τῆς ἀνατολῆς *rising azimuth* of sun or star, counted from the east; *cf.* appendix 11. Copernicus, *De revol.* II,7: *latitudo ortus*.

πλάτος τοῦ κύκλου τῆς κινήσεως[6] an astrological concept based on the following relation. Let a planet be at a distance d from upper (or lower) culmination, c the length of half the arc of daylight (or night) and ϕ the geographical latitude. The "*latitude of the circle of motion*" p is defined by $\frac{p}{\phi} = \frac{d}{c}$.

[1] fol. 267ᵛ,25 to 27; 270ʳ,32.
[2] fol. 412ᵛ, 413ʳ; 420ᵛ, 421ʳ.
[3] fol. 413ᵛ, 414ʳ; 421ᵛ, 422ʳ.
[4] fol. 323ᵛ,31; 338ᵛ. For the origin of this terminology *cf.* Nallino, Batt. II, p. 339.
[5] fol. 265ʳ,19.
[6] fol. 312ʳ,1 to 10 (*Sanjarī zīj*, Book XII, Ch. 2, 2).

πλέον καὶ ἔλαττον

increment; equivalent: ἐκτλῆφι.

for the moon:

increment of the second lunar equation over the first; *cf.* appendix 6 and fig. 9*b*.

for the planets:

similar; *cf.* appendix 7. *Cf. also* ἀνώμαλος.

πλέον καὶ ἔλαττον ὄψεως[1] or πλέον καὶ ἔλαττον ἀπὸ τῆς ὄψεως[2] *parallax*.

combined with μῆκος or πλάτος:[2] longitudinal or latitudinal component of parallax.

πλέον ἔλαττον ἰδίου[3] parallax as function of anomaly; *cf.* appendix 10.

[1] tables: fol. 360ᵛ to 361ᵛ.
[2] fol. 360ᵛ.
[3] fol. 358ᵛ, col. 12.

πόρρω μῆκος *see* μῆκος ἐπιμηκέστερον

ῥοέτ

vision (*cf.* Nallino, Batt. II, p. 334); *see* ντατὶλ σαὰτ ῥοέτ. *Cf. also* CCAG 1, p. 88, 28.

σαάτ

hour; *see* ντατὶλ σαὰτ ῥοέτ and ἐσσὰ σαὰτ μπότ.

σαγίτα

sagitta[1] $= R - \text{Sin}(90 - \theta) = R(1 - \cos\theta)$ for $R = 60$.

ἡ σαγίτα δὲ ἡ μεγάλη διάμετρος ἐστὶ τοῦ κύκλου[2] *diameter*.
σαγίτα τῆς ἡμέρας[3] *sagitta of the ascensional difference* (the ὄρθωσις τῶν ἡμερῶν).

[1] Definition: fol. 283ᵛ,16 f.; tabulated: fol. 432ᵛ; figure: Vat. gr. 211, fol. 115ʳ. For the translation of Latin astronomical works into Greek *cf.*, e.g., Vat. gr. 212 (Cod. Vat. Gr. I, p. 270 ff.).
[2] fol. 283ʳ,31.
[3] fol. 285ᵛ,19.

σημεῖον

sign, indicator for the combination of cases; e.g., a coefficient c is given the σημεῖον $\bar{\alpha}$ if its argument belongs to the first column of a table, $\bar{\beta}$ if it belongs to the second column. Similarly the preliminary latitude β'_3 (of Venus) obtains a σημεῖον $\bar{\alpha}$ or $\bar{\beta}$ depending on its argument. The combinations of the two σημεῖα then decide whether the latitude $c\beta'_3$ is northern or southern[1] (*cf.* appendix 8).

σημεῖον τῆς μοίρας τῆς ἀναβάσεως
ὄρθωσις σημείου
τραχηλαία τοῦ σημείου
} *see* appendix 14.

[1] fol. 297ᵛ,11 f.: ἐὰν τὰ δύο σημεῖα ἐξισοῦνται τὸ πλάτος εἰς τὸ βόρειον μέρος· εἰ δ' οὐκ ἐξισοῦνται τὸ πλάτος εἰς τὸ νότιον.

σκιά

σκιὰ ἐκλείψεως ☉ [1] *eclipse magnitude* for the sun.

Shadow, used as technical term for tan α and its modifications:

σκιὰ τῆς πρώτης μετακλίσεως [2] Tan δ ($R = 60$, $\delta =$ declination).

σκιὰ τῶν δακτύλων [3] $12 \cdot \cot \alpha$.

σκιὰ τῶν ποδῶν [3] $7 \cdot \cot \alpha$.

[1] fol. 428v.
[2] fol. 431r.
[3] fol. 431v, $\alpha =$ noon altitude of the sun (ἀνάβασις ☉).

σκιάσμα

σκιάσμα τῶν δακτύλων [1] or σκιὰ δακτύλων [2] $12 \cdot \cot \theta$.

σκιάσμα or σκιά [3] Tan α ($R = 60$).

ὄρθωσις σκιάσματος *see* ὄρθωσις.

[1] fol. 365r, main title.
[2] fol. 365v, column title.
[3] fol. 283r,29; 431v.

σκοτεινός

οἱ ἀστέρες δὲ οἱ ἀμυδρῶς φαινόμενοι, τούτους ὁ πτολεμαῖος ἀστέρας σκοτεινοὺς ἐκάλεσεν [1] "the faintly visible stars which Ptolemy called *obscure*" (in fact, however, he called them νεφελοειδεῖς [2] *nebulous*).

[1] fol. 317r,3 f.
[2] Almag. VIII,1 (Heib. II, p. 169,18).

στάσις *see* ὥρα

στερεός

στερεὰ ὄρθωσις τοῦ ἡλίου [1] title of a table of true solar longitudes (counted from ♈ 0°) as function of the mean longitude, counted from the apogee. Thus we have here the sum of the mean longitude and the solar equation denoted by a term which ordinarily would be used only for the equation; *cf.* ὄρθωσις and ὄρθωμα.

[1] fol. 378v, 379r.

σφαῖρα

λεπτὰ σφαίρας διαταγωγῆς [1] coefficients, depending on the lunar velocity, increasing from 0 to 0;12, probably indicating the increase of the diameter of the shadow.

[1] fol. 428v, last column. Same heading in Vat. gr. 211, fol. 229v.

τακκαέκ

τακκαὲκ ἐσσὰς κατὰ πέρσας ἤτοι μερισμὸς τῶν λεπτῶν [1] from *daqā'iq-i ḥiṣaṣ* [Kennedy]; *cf.* Nallino, Batt. II, p. 328. Coefficient of interpolation for planetary latitudes (Almag. XIII,5: ἑξηκοστά; Heib. II, p. 582).

[1] fol. 398v/399r; *cf.* μερισμός.

τασιρὴν ἀουάλ

ἔπειτα τηρεῖται ἡ ἀρχὴ τοῦ χρόνου ἤτοι ἡ ἀρχὴ τοῦ τασιρὴν ἀουὰλ ποία ἡμέρα ἔστιν ἀπὸ τῶν τῆς ἑβδομάδος [1] "then one has to determine which day of the week is the first day of the year." From *tasyīr-i awwal* [Kennedy].

[1] fol. 333v, right table 6 f.

τέλειος

final value of some parameter, e.g., τέλειον ὕψωμα or ὄρθωσις τέλεια; *cf.* appendix 1.

see also: ἀνάβασις; ἀποκατάστασις; δάκτυλος; διακρίνω; πλάτος.

δάκτυλος τέλειος *area digit* of eclipse magnitudes; *cf.* δάκτυλος.

τέλειος κύκλος τῆς ἡμέρας [1] or ἡ τελεία τῆς ἡμέρας ζώνη [2] celestial *equator.*

[1] fol. 285r,3.
[2] fol. 284r,3.

τελῶ

τόξον τετελειωμένον [1] *complement of an arc,* i.e., $90 - \theta$; similar for all specific angular distances, e.g., τετελειωμένη ἀνάβασις [2] *complement of altitude.*

ἡ τραχηλαία ἡ τετελειωμένη [3] Sin of the complement.

[1] fol. 283r,9.
[2] fol. 299v,18.
[3] fol. 299r,10.

τηρῶ

τηρεῖται εἰς τὰ ιη' καὶ τὸ ἐξελθὸν μερίζεται εἰς τὰ ιζ' [1] "*multiply* by 18 and divide the result by 17."

[1] fol. 300r,12 f.

τόξον

arc

τόξον τετελειωμένον [1] *complement of an arc,* i.e. $90 - \theta$.

τόξον τῆς θεωρίας *arc of vision*; for the moon *cf.* appendix 4; for the planets: *elongation* necessary for visibility.[2]

τόξον τῆς ἡμέρας [3] *length of daylight* (measured in degrees).

τόξον τοῦ φωτός *see* φῶς.

[1] fol. 283r,8.
[2] fol. 309v,17; tabulated for clima 4: fol. 359r.
[3] fol. 285v,29.

τόπος

place, longitude; equivalent: μῆκος.

see also: ἄκρον; θεμέλιον; ὄψις.

τόπος τῆς τύχης τῶν ζῳδίων [1] *rising time* or *oblique ascension,* counted from ♑ 0° as is customary in the Handy Tables of Ptolemy and Theon; by adding 90° one obtains the rising times counted from ♈ 0°. *Cf. also* appendix 4.

I have no suggestion to offer for the explanation of the name "*place of fortune*" for what is commonly called ἀναφοραί, i.e. *rising times.* There is no relation to the astrological terms κλῆρος τῆς τύχης " Lot of Fortune " or ἀγαθὴ and κακὴ τύχη " Good " and " Bad Fortune " (i.e. 5th and 6th locus).[2]

κανόνιον τοῦ τόπου τῆς τύχης [3] *table of rising times*, counted from ♈ 0°.

τόπος τῆς τύχης τῶν ζωδίων μετὰ τῆς εὐθείας γραμμῆς [4] *right ascension* (also μῆκος is used,[5] instead of τόπος).

τόπος τῆς τύχης τοῦ αὐθημερινοῦ τοῦ ☉ μετὰ τοῦ ἀπὸ τοῦ πλάτους τῆς πόλεως [6] *oblique ascension* of $\lambda_{\odot}$.

τόπος τῆς τύχης τῆς τύχης τοῦ πλάτους τῆς πόλεως [7] *oblique ascension of the ascendant* at a given place (also the 10th locus seems possible). Note the duplication of τῆς τύχης which is part of the terminology.

[1] fol. 284v,11 to 29 (*Sanjarī zīj*, Book III, ch. 4); also fol. 286r,13 to 18 (*loc. cit.* Book IV, ch. 4).
[2] For these concepts *cf.* Bouché-Leclercq, AG p. 289 f. and p. 280.
[3] fol. 265r,25 f., 28; tables: fol. 433v to 438v.
[4] fol. 284v,12,16; 286r,15. Tabulated: Vat. gr. 211 fol. 157r.
[5] fol. 299v,3.
[6] fol. 289r,11 f.
[7] fol. 289r,12; similarly fol. 289r,29: τόπος τῆς τύχης τῆς τύχης.

τραχηλαία

$\text{Sin}\,\theta = R \sin\theta$ for $R = 60$.[1]

τραχηλαία τῆς ὀρθώσεως τῶν ἡμέρων Sin a where $a =$ τόξον τῆς ὀρθώσεως τῶν ἡμερῶν.[2]

μεγάλη τραχηλαία ἥμισύ ἐστι τῆς διαμέτρου [3] *sinus totus* = *radius*.

τραχηλαία τοῦ σημείου *see* appendix 14.

[1] tabulated fol. 365v and 432r for single degrees.
[2] fol. 366v; similar fol. 373v.
[3] fol. 283r,20 f.; also Vat. gr. 211 fol. 115r, figure.

τρυτάνη

τρυτάνη τῆς θεωρίας τῆς ☾ [1] *visibility limit* for the new crescent.

τρυτάνη in connection with the ιβ′ οἰκήματα [2]: " *balancing* " of the length of the loci at sphaera recta and for given latitude, i.e. comparison of their relative length.

[1] fol. 308r,22.
[2] fol. 289v,13.

τύχη

τύχη τοῦ καιροῦ ἐκείνου [1] or τύχη τοῦ καιροῦ [2] the *ascendant* at the given moment.

τόπος τῆς τύχης and τόπος τῆς τύχης τῆς τύχης *see* τόπος.

[1] fol. 304v,16.
[2] fol. 299r,9.

ὕψωμα

apogee; also *motion of apogee*; *passim*. *Cf.* appendix 1.

ἀτελὲς ὕψωμα *see* ἀτελής.

ὕψωμα τοῦ ☉ εἰς τὸ μέσον τῆς ἡμέρας [1] *noon altitude* of the sun.

[1] fol. 265r,21.

φῶς

τόξον τοῦ φωτὸς ἤγουν τῆς ἐλλάμψεως τῆς ☾ [1] *arc of light* or illumination of the moon, i.e. ripeness of the crescent, defined by $\sqrt{\beta^2 + \Delta\lambda^2}$ where $\beta =$ lunar latitude, $\Delta\lambda$ the elongation from the sun.

[1] fol. 307v,3 f.

χασσᾶ

χασσᾶ μαντὰλ ἤτοι ἰδία κίνησις [1] *equated argument* (of anomaly). From Persian *khāṣṣa mu'addal*; *cf.* Bīrūnī, Astrol. § 183, p. 94; Nallino, Batt. I, p. 213 (2), II p. 329; Sédillot, Tables astron., p. 45 ff.

τοῦ χατζᾶ ἤγουν τῆς ἰδίας κινήσεως [2] *anomaly*.

[1] fol. 263r,28.
[2] fol. 264r,26.

χθαμαλοὶ ἀστέρες

the *inferior planets*.[1]

[1] fol. 422v.

χουσούφ

lunar eclipse: κουσοὺφ λέγεται κατὰ πέρσας ἤτοι ἔκλειψις· τῆς δὲ ☾ χουσοὺφ ἤτοι κρύψις [1] from Arabic *kusūf* and *khusūf* respectively [Kennedy].

[1] fol. 269v,9 f.

χρόνος

year, of any particular type.

χρόνοι ἀράβων [1] *Arabic* (lunar) *years* (of 354 or 355 days) of the era Hijra (A.D. 622 July 16 or 15).

χρόνοι περσῶν [1] *Persian years* (of 365 days) of the era Yazdigerd (A.D. 632 June 16).

χρόνος τοῦ ☉ [2] *solar year*, i.e. tropical year.

[1] fol. 333r; 376r.
[2] fol. 310v,11.

ψῆφος

number—passim.

ψῆφος μέσος [1] calendaric computation with months of either 29 or 30 days, arranged in fixed order, in contrast to a true lunar calendar which is based on the first visibility of the crescent.

[1] fol. 278r,9.

ὥρα

κίνημα τῶν ὡρῶν *see* κίνημα.

μερισμοὶ τοῦ μήκους τῶν ὡρῶν *see* ἐσσὰ σαὰτ μπότ.

ὀρθὴ ὥρα *see* ὀρθός.

ὥρα τῆς ἀποκαταστάσεως *see* ἀποκατάστασις.

ὥρα τῆς μήκους *see* μῆκος.

ὥρα πεσοῦσα *half duration* of an eclipse, i.e. from first contact to eclipse middle; for lunar[1] and for solar[2] eclipses. *Cf.* appendix 9.

ὥρα στάσεως[3] *half duration of totality* of a lunar eclipse; also simply στάσις.[4] Equivalent: ὥρα τοῦ καιροῦ.[5]

ὥρα τῆς ἡμέρας (or τῆς νυκτὸς) πάσης[6] *length of daylight* (or of night).

ὥρα τοῦ μέσου τῆς ἡμέρας[7] *half length of daylight*; *cf.* appendix 9.

ὧραι μήκους τῆς ἡμέρας[8] *longest daylight.*

[1] fol. 359^v, tabulated for the moon in apogee; also fol. 427^r, 428^v, tabulated for four values of the lunar velocity.
[2] fol. 362^r, left table; for the moon in apogee.
[3] fol. 359^v.
[4] fol. 427^r.
[5] fol. 269^r,18,20.
[6] fol. 268^v,27 f. and fol. 269^r,24 respectively.
[7] fol. 270^r,14 f.
[8] fol. 360^v to 361^v.

ὡραῖον κανόνιον

proper table, i.e. a table obtained by interpolation between existing tables.[1]

[1] For parallax at a given geographical latitude and for a given degree within a zodiacal sign; fol. 271^v and 304^r.

APPENDICES

APPENDIX 1

Example for finding the solar longitude for A.H. 695 VI 10[1] (= A.D. 1296 April 15). From the tables:[2]

	μέση κίνησις	ὕψωμα
for 691 years	9^s 10;25,56°	10;8,37°
4 years	10 16;40, 1	0;3,32
6 months	4 25;52,35	0;0,22
10 days	0 8;52,15	0;0, 1
total:	1^s 1;50,47	10;12,32
	θεμέλιον:	2^s 18;20
	τέλειον ὕψωμα τοῦ ☉:	2^s 28;32,32
	μέση κίνησις:	13^s 1;50,47
	κέντρον:	10^s 3;18,15

With the κέντρον as argument, from the tables called κανόνια τῆς ὀρθώσεως τοῦ ἡλίου:[3]

	ὄρθωσις:	περίσσεια
for 10^s 3°:	1;49, 0°	—0;1,16°
0;18,15°:	—0; 0,18	(from —0;1,16 · 0;18,15 ≈ —0;1 · 0;18)[4]
ὄρθωσις τελεία:	1;48,42	
κέντρον:	10^s 3;18,15	
ὕψωμα:	2^s 28;32,32	
	1^s 3;39,29[5]	τὸ αὐθημερινὸν τοῦ ☉.

Explanation of this Procedure (*cf.* fig. 3)

From the tables for 30 Arabic years, single years, months, and days, is found the day of the week, the mean longitude $\bar{\lambda}$ (μέση κίνησις) of the sun, and the motion v of the solar apogee since epoch (ὕψωμα):

$\bar{\lambda} = 1^s\ 1;50,47°$ $\qquad v = 10;12,32°$

apogee at epoch: $A_0 = 2^s\ 18;20$ θεμέλιον

thus present apogee: $\lambda_A = 2^s\ 28;32,32$.

Thus we have the anomaly (κέντρον):

(1) $\qquad a = \bar{\lambda} - \lambda_A = 10^s\ 3;18,15$

to which belongs, according to the tables for the equation, a correction of

$$\delta = +\ 1;48,42$$

and therefore a true longitude (αὐθημερινόν) of

(2) $\qquad \lambda = \bar{\lambda} + \delta = 1^s\ 3;39,29°$.

Instead of this simple addition, the text uses $\lambda = \delta + a + \lambda_A$ which is also correct because of (1).

[1] fol. 322^v,6 to 323^r,8.

[2] fol. 334^v to 338^r. The text also contains a check for the day of the week.

[3] fol. 339^v.

[4] With a less crude rounding one would obtain 0;0,23.

[5] Text incorrectly 3;40,29.

APPENDIX 2

Example for finding the longitude of the moon for A.H. 652 IV 22[1] (= A.D. 1254 June 11). From the tables:[2]

	μέση κίνησις	ἰδία κίνησις	κέντρον ☾
for 631 years	7^s 19;16°	6^s 2;17°	1^s 21;28°
21 years	4 18;45	0 29;36	0 7; 5
4 months	3 2;42	2 22;47	0 9;57
22 days	9 6;42	9 4;22	5 2; 1
total:	0^s 17;25	6^s 29; 2	7^s 10;31

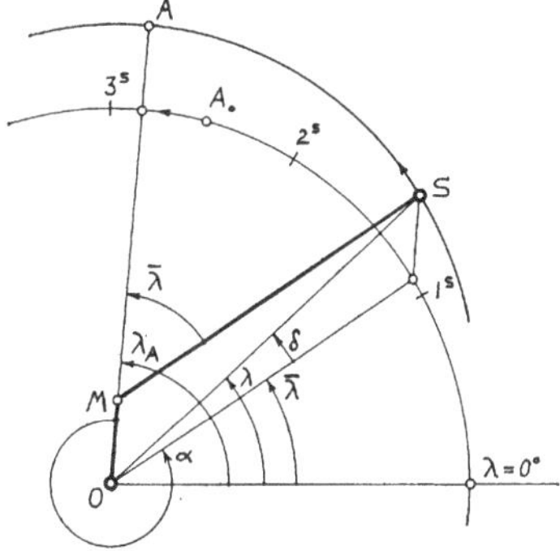

FIG. 3

Correction for geographical longitude 87° from 90°, i.e. for a difference of 3° = 0;12^h:[3]

for 0; 2^h	0;1, 5,53°	0;1, 5,20°	0; 2, 1,55°
0;10^h	0;5,29,24[4]	0;5,26,37	0;10, 9,33
0;12^h	0;6,35,17	0;6,31,57	0;12,11,28
	≈ 0; 7	≈ 0;7	≈ 0;12
previous results:	17;25	6^s 29;2	7^s 10;31
for long. 87°:	0^s 17;32	6^s 29;9	7^s 10;43

Entering with the κέντρον as argument the tables called κανόνια τῆς ὀρθώσεως τῆς σελήνης[5] one finds[6] in column γ:

	ὄρθωσις ᾱ	περίσσεια
for 7^s 10°:	11;11	+0;9
0;43:	+ 0;6	(from 0;9 · 0;43 = 0;6,27)
ὄρθωσις α′:	—11;17	
ἰδία:	6^s 29; 9	
ἴδιον τέλειον:	6^s 17;52 = χασσᾶ μαντάλ.	

With the *ἴδιον τέλειον* as argument one finds in the *κανόνια τῆς ὀρθώσεως τῆς σελήνης*[7] in column *ε*:

	ὄρθωσις $\bar{\beta}$	*περίσσεια*
for $6^s\,17°$:	1;37	+ 0;5
0;52:	+ 0; 4	(from 0;5 · 0;52 = 0;4,20)
ὄρθωσις β':	1;41	

With the *κέντρον* $7^s\,10;43 \approx 7^s\,11°$ as argument, one finds in the same tables[6] in column *δ*:

γενικὰ λεπτά: 0;51.

With the *ἴδιον* (*τέλειον*) from column *ϛ*:[7]

	μῆκος	*περίσσεια*
$6^s\,17°$:	0;59	—0;3
0;52:	—0; 3	(from —0;3 · 0;52 = —0;2,36)
	0;56	

Multiplying this result with the coefficient obtained as *γενικὰ λεπτά* one finds:

$$0;56 \cdot 0;51 = 0;47,36 \approx 0;48$$

called *μῆκος ἐγγύτερον*. To this quantity is then added the *ὄρθωσις δευτέρα*

$$0;48 + 1;41 = 2;29 = \text{ὄρθωσις στερεά}$$

which is (incorrectly) taken as a negative correction of the *μέση κίνησις*

$$0^s\,17;32° - 2;29 = 0^s\,15;3°.$$

This is the final result, the *αὐθημερινὸν τῆς* ☾. The correct result would have been 17;32 + 2;29 = 20;1°.

Explanation of this Procedure (*cf.* fig. 4)

The method for finding the true longitude of the moon is based on the model of the lunar motion devised

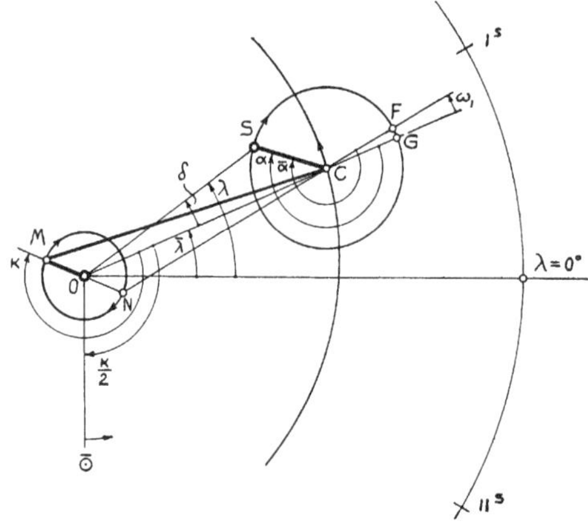

FIG. 4

by Ptolemy in Book V of the Almagest. Consequently, one determines first the mean longitude ($\bar{\lambda}$ = *μέση κίνησις*), the mean anomaly ($\bar{a}$ = *ἰδία κίνησις*, counted from the mean apogee F), and the double elongation (κ = *κέντρον*) for the moment in question, including the small addition due to the change from longitude 90° to the longitude 87° of Buchara.

With κ as argument one finds the "first correction" (ω_1 = *ὄρθωσις* a') which gives the angular distance of the "true apogee" G of the epicycle from the "mean apogee" F. By subtracting it from $\bar{a}$ one obtains the true anomaly a (*ἴδιον τέλειον*), that is, the angular distance of the moon on the epicycle from the true apogee as seen from O.

With a as argument one can now find as "second correction" (ω_2 = *ὄρθωσις δευτέρα*) the epicyclic equation of the moon under the assumption that the epicycle is at its maximum distance from O. This is the value one would obtain if the deferent of the moon were at maximum distance. If the epicycle is, on the contrary, at its minimum distance, the same true anomaly a would lead to a greater equation. The corresponding increment μ is called *μῆκος*. In general, however, the epicycle will be in an intermediate position, i.e. the double elongation κ is neither 0° nor 180°. The "proportional parts" (*γενικὰ λεπτά*) tabulated as function of κ give the sexagesimal fraction γ of μ that corresponds to an intermediate distance. Thus the increment of the epicyclic equation is only $\gamma \cdot \mu$, called *μῆκος ἐγγύτερον* "nearer distance." Consequently, the total equation (*ὄρθωσις στερεά*) is given by

$$\delta = \omega_2 + \gamma \cdot \mu$$

and thus the true longitude (*αὐθημερινόν*) by

$$\lambda = \bar{\lambda} + \delta.$$

[1] fol. 324ᵛ,12 to 325ʳ,31.
[2] fol. 334ᵛ to 338ʳ.
[3] fol. 336ᵛ.
[4] This is the correct value; the table itself (fol. 336ᵛ) has here a scribal error (5,25,24).
[5] fol. 340ᵛ to 343ʳ.
[6] fol. 342ᵛ.
[7] fol. 343ʳ.

APPENDIX 3

Example for finding the longitude of Saturn for A.H. 652 IV 22[1] (= A.D. 1254 June 11). From the tables for mean motion[2] one finds

μέση κίνησις: $9^s\,18;35°$ *ἰδία κίνησις*: $5^s\,7;54°$[3]

and similarly the

ἀτελὲς ὕψωμα τοῦ ♂:	$0^s\ 9;34,29°$.[4]
To this is added the *ὕψωμα τοῦ* ♄:	$8^s\ 0;45$[5]
which results in the *ὕψωμα τέλειον*:	$8^s\,10;19$.

This longitude is subtracted from the

	μέση κίνησις:	$9^s\ 18;35°$
resulting in the	κέντρον:	$1^s\ 8;16.$

With the κέντρον as argument one enters the tables called κανόνια τῆς ὀρθώσεως τοῦ κρόνου.[6] There one finds in column γ

for $1^s\ 8°$:	ὄρθωσις $\bar{\alpha}$:	3;50	περίσσεια: + 0;5
0;16:	:	+ 0; 1	(from $0;5 \cdot 0;16 = 0;1,20$)
thus ὄρθωσις $\bar{\alpha}$:		3;51.	

Subtracting this result from the κέντρον, but adding it to the κίνησις, one finds

	$1^s\ 8;16$		$5^s\ 7;54$
	$-\ 3;51$		$+\ 3;51$
κέντρον τέλειον:	$1^s\ 4;25°$	ἴδιον τέλειον:	$5^s\ 11;45°$

Entering the tables with the first of these values, one finds in column δ the

γενικὰ λεπτά = 0;49 [7]

and similarly [8] with the second value from column ϛ:

for $5^s\ 11°$:	ὄρθωσις $\bar{\beta}$:	2;16	περίσσεια: $-\ 0;7$
0;45:		$-\ 0;\ 5$	(from $-\ 0;7 \cdot 0;45 = -\ 0;5,15$)
thus ὄρθωσις δευτέρα τελεία:		2;11°.	

Once more with the ἴδιον τέλειον as argument we enter the same tables and find in column ε [9]

for $5^s\ 11°$: πόρρω μῆκος: 0;8 περίσσεια: 0;0.

This quantity is now multiplied by the γενικὰ λεπτά

$$0;8 \cdot 0;49 = 0;6,32 \approx 0;7$$

and the result, which is called μουδρούπ ἤτοι μίγμα, is subtracted from the

	ὄρθωσις δευτέρα τελεία:	2;11
		$-\ 0;\ 7$
resulting in the	ὄρθωσις στερεά:	2; 4.
To this we add the	κέντρον τέλειον:	$1^s\ 4;25$
and obtain the	ὄρθωσις τετελειωμένη:	$1^s\ 6;29.$
This amount is added to the	ὕψωμα τέλειον:	$8^s\ 10;19$
producing the final result as the αὐθημερινὸν τοῦ ♄.		$9^s\ 16;48°$

Explanation of this Procedure (*cf.* fig. 5)

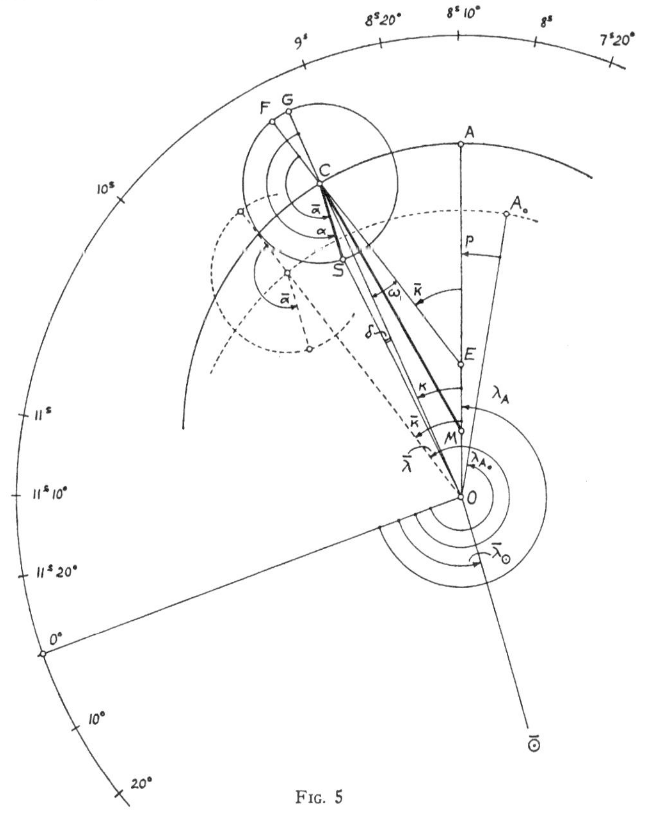

FIG. 5

The method is again essentially Ptolemaic (Almagest Book XI, 11, 12). One first finds from the tables of mean motion the mean longitude $\bar{\lambda}$ (μέση κίνησις) and mean anomaly $\bar{a}$ (ἰδία κίνησις) represented in fig. 5 on a concentric deferent (dotted circles). Then one determines the position of the apogee of Saturn for the given moment by adding the motion of precession since epoch, p, to the longitude $\lambda_{A_0} = 8^s\,0;45°$ of the apogee at epoch:

$$\lambda_A = \lambda_{A_0} + p.$$

The difference

$$\bar{\kappa} = \bar{\lambda} - \lambda_A$$

called κέντρον, gives the distance of the center C of the epicycle from the apogee for the given moment. According to the Ptolemaic theory, however, this angle has not to be measured at O but has its vertex in E, the "equant" in later terminology. The actual center of the deferent is the midpoint M between E and O.

The epicycle with center C carries the planet S, which has a mean anomaly $\bar{a}$, counted from the mean apogee F, that is to say, from the apogee that corresponds to the equant. In order to find the epicyclic equation as seen from O one has to know the distance of S from the true apogee G, i.e., the true anomaly a (called here ἴδιον τέλειον). This difference

$$\omega_1 = a - \bar{a}$$

is tabulated as function of κ and called "first correction." The same angle ω_1 gives also the difference between the κέντρον $\bar{\kappa}$ which increases proportionately with time, and the "final center" κ which measures the distance of C from A as seen from O:

$$\kappa = \bar{\kappa} - \omega_1.$$

We now know the distance of C from A and the position of S on the epicycle with respect to the true apogee G. This would immediately lead to the equation δ which is the difference between the longitude $\bar{\lambda}$ of C and λ of S, if the epicycle were at mean distance from O. The corresponding values, called "second correction" ω_2, are tabulated as function of a in the sixth column of our tables. In fact, however, the epicycle is nearer to the apogee. If it were exactly in the apogee, the equation would be less than ω_2 by the amount μ_A, tabulated in column 5 for the same entry a.[10] Actually, however, C is not in A, but at a distance κ from it. Therefore, one finds in column 4 coefficients γ (called γενικὰ λεπτά) as function of κ. Hence the final equation is given by

$$\delta = \omega_2 - \gamma \cdot \mu_A$$

and thus the true longitude of the planet by

$$\lambda = \lambda_A + \kappa + \delta.$$

[1] fol. 325ᵛ,1 to 326ʳ,13.
[2] fol. 334ᵛ to 338ʳ.
[3] Only this result is given in the text (fol. 325ᵛ,5 f.) but it can be easily checked by means of the same operation, described at the beginning of appendix 2. It turns out that the first figure should be $9^s\,19;15°$ while the second is correct.
[4] This figure is correct. The single contributions are listed in the last columns of the tables fol. 335ʳ to 338ʳ under the heading κίνησις ὑψώματος. κίνησις τῶν ἀπλανῶν ἀστέρων. Thus all apsidal lines are considered to have fixed sidereal longitudes.
[5] From fol. 336ᵛ.
[6] fol. 343ᵛ to 346ʳ.
[7] It suffices to use the entry $1^s\,4°$ (fol. 344ʳ).
[8] fol. 346ʳ.
[9] called μῆκος ἐπιμηκέστερον.
[10] In the opposite case of C in the perigee, ω_2 would increase by μ_P tabulated as function of a in column 7.

APPENDIX 4

Example for deciding whether or not the new moon will be visible on the first day of Ramadan A.H. 701 (= A.D. 1302 April 30). A crude method for answering this question is described fol. 331ʳ, 6-30. It consists in the following steps:

One computes for the preceding day (i.e. Sha'bān 29) the true longitude (αὐθημερινόν) for sun, moon, and ascending node. The result[1] is

$$\lambda_{\odot} = 1^s\,16;56$$
$$\lambda_{☾} = 1^s\,27;21$$
$$☊ = 9^s\,28;40.$$

Then one adds 180° to the longitudes of sun and moon and enters with the result the table for rising times or oblique ascensions for the given climate (i.e. Byzantium). These tables are called in the text τὰ κανόνια τοῦ τόπου τῆς τύχης τῶν ζωδίων[2] but are headed (fol. 434ʳ to 438ᵛ) κανόνιον τῶν ἀναφορῶν τῶν ζωδίων. One finds[3]

$$\bar{\rho}_{\odot} = 240;59 \qquad \bar{\rho}_{☾} = 254;48$$

and from it

$$\Delta\bar{\rho} = \bar{\rho}_{☾} - \bar{\rho}_{\odot} = 13;49.$$

This difference is called μῆκος μέσον τοῦ ☉ καὶ τῆς ☾ (i.e. literally "distance between sun and moon" but actually representing the oblique ascension of the elongation) or also τὸ τόξον τῆς θεωρίας i.e. "arc of vision"[4] (fig. 6).

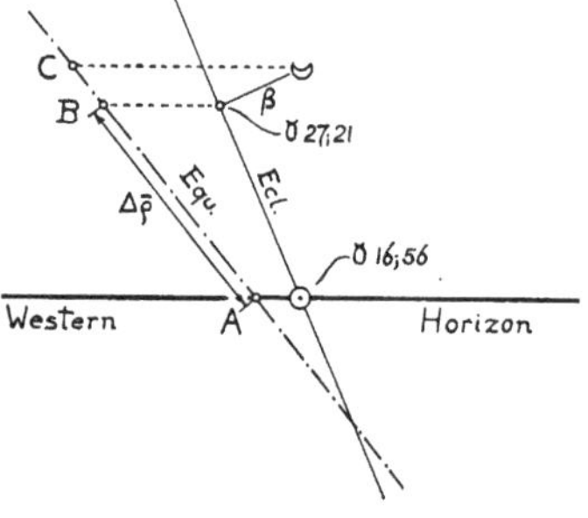

FIG. 6

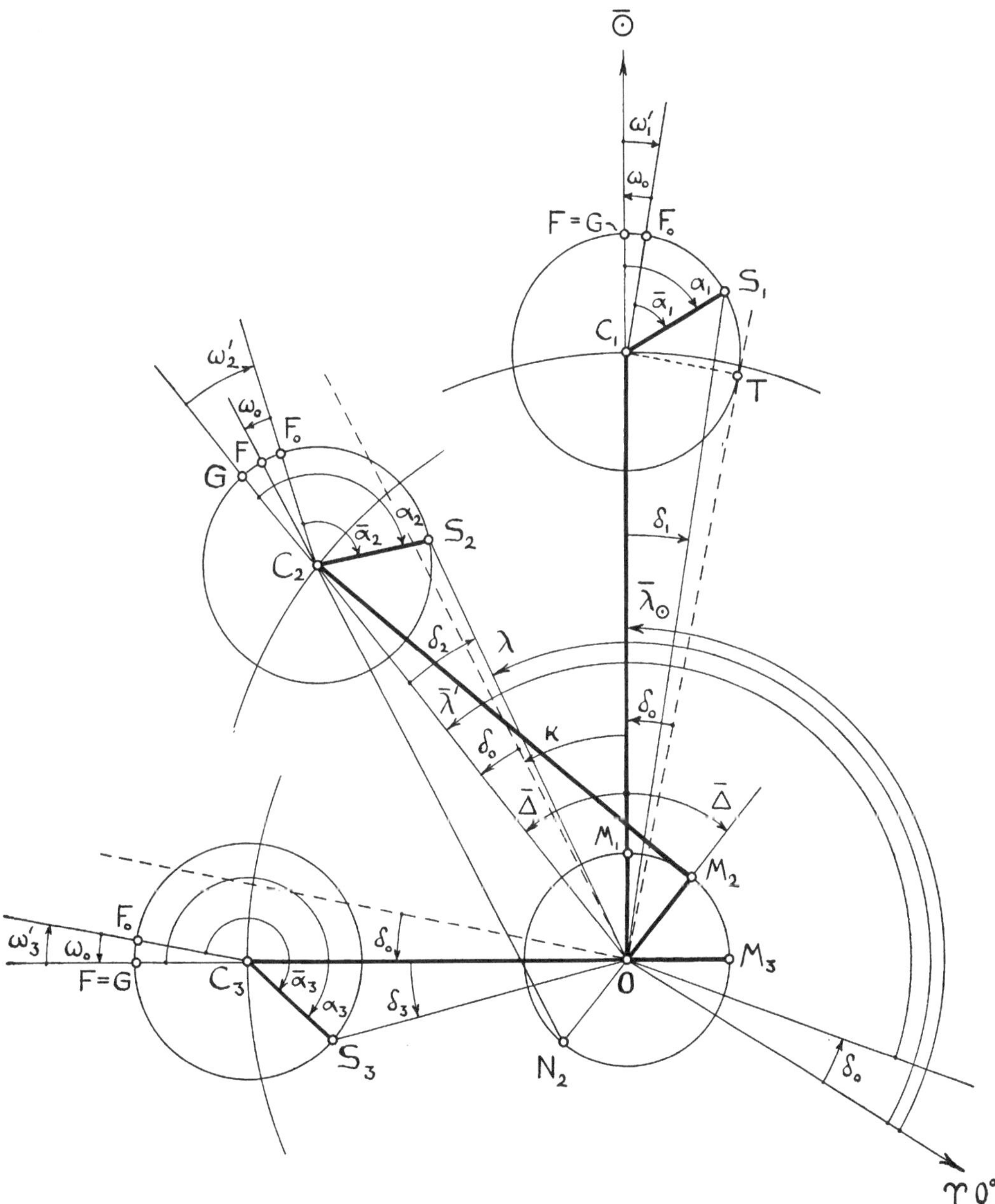

FIG. 7

In the next step one finds the argument of latitude from

$$\omega = \lambda_☾ - ☊ = 1^s\,27;21 - 9^s\,28;40 = 3^s\,28;41$$

and with ω as argument in the tables of latitudes

$$\beta = +4;25.^5$$

Finally one computes

$$\Delta\bar{\rho} + \tfrac{1}{2}\beta = 13;49 + 2;12 = 16;1.$$

Since 16;1 > 12 the moon will be visible on the evening of the 29th. A result < 10 would have meant invisibility; obviously the method is not considered reliable enough to decide the cases between 10 and 12.

Fig. 6 explains the underlying idea. The oblique ascension AB = $\Delta\bar{\rho}$ measures the time between sunset and the setting of the point which has the same longitude as the moon. The correction BC for a lunar latitude β is then simply approximated by $\frac{1}{2}\beta$, positive for northern latitudes, negative for southern.

[1] Including a correction for geographical longitude.
[2] fol. 331r,13 f.
[3] Interpolation results actually in 240;59,48 and 254;31,57 respectively.
[4] fol. 331r,20 f. and 25. This is not the "arcus visionis" of modern terminology.
[5] From fol. 431r for $\omega = 3^s\,28°$. For $\omega = 3^s\,28;41$ one would obtain 4;23,40.

APPENDIX 5

Epoch values (θεμέλια) for the beginning of the year Yazdigerd 541 I 1 (= A.D. 1172 Feb. 2):

sun: mean longitude from apogee:
$7^s\,20;58,58,43$ μέση κίνησις [1]
longitude of apogee:
$2^s\,27;50,42,36$ ὕψωμα [2]
thus mean longitude:
$10^s\,18;49,41,19$ ☉ κίνησις [3]

moon: mean longitude:
$0^s\,26;50,29,46$ μέση κίνησις [4]
thus mean elongation:
$2^s\,8;\,0,48,27$ μετάβασις [5] or κέντρον.[6]

[1] fol. 376r; fol. 377r for Yazd. 541: $7^s\,20;58,59$.
[2] fol. 376r marg. note.
[3] fol. 375r.
[4] fol. 376r; fol. 379v for Yazd. 541: $0^s\,26;50,30$.
[5] fol. 376r.
[6] fol. 379v for Yazd. 541: $2^s\,8;0,48$.

APPENDIX 6

The examples discussed in the appendices 1 to 3 concern the tables which occupy fol. 334v to 358r. These tables are based on Arabic years and normed for a geographical longitude of 90°; the technique of computation is practically identical with the procedures of the Almagest.

Our manuscript contains also a second set of tables (fol. 375r to 422r) which is normed for a geographical longitude of 84°, for the epoch Yazdigerd 541 I 1 (= A.D. 1172 Feb. 2; *cf.* appendix 5) and uses Persian years. The most essential deviation from the previous tables consists, however, in the fact that all corrections of the mean positions are arranged in such a fashion that they are never negative.[1] This constitutes a great practical advantage over the Ptolemaic fashion. According to a remark on fol. 261r,14 these tables are from the *zīj al-'Alā'ī* (σύνταξις ἀλαῆ).[2]

We give in the following a detailed description of the lunar tables. The planetary tables follow exactly the same idea (*cf.* appendix 7) and the terminology is therefore the same.

Fig. 7 gives a schematic description (not drawn to scale) for three typical cases in the lunar motion:

(1) means conjunction, i.e., the observer O, the center C_1 of the lunar epicycle, and the mean sun are in a straight line

(2) at elongation $\bar{\Delta} = \kappa + \delta_0$ of the mean moon (= center C_2 of the epicycle) from the mean sun

(3) at quadrature, i.e., at elongation of $\bar{\Delta} = 90°$ or $\kappa = 85°$.

Notation: O observer

- M movable center of eccenter, moving on the circle of radius OM = e in retrograde direction, by an amount which equals the elongation $\bar{\Delta}$
- N point diametrically opposite M on circle of radius e
- C center of epicycle = "mean moon," at elongation $\bar{\Delta}$ from mean sun
- TO tangent to epicycle when in apogee of eccenter
- δ_0 apparent radius of epicycle when in apogee of eccenter, seen from O ($\delta_0 = 5;0°$ on fol. 383v)
- $\bar{\lambda}'$ "mean longitude," increasing linearly with time, counted from ♈0° + δ_0
- G "true apogee" of epicycle, defined by straight line OCG
- F "mean apogee" of epicycle, defined by straight line NCF; maximum angular distance possible between F and G $\omega_0 = 14;37°$ (on fol. 382v)
- F_0 point on the epicycle such that $FF_0 = \omega_0$
- S moon on epicycle
- $\bar{a}'$ "mean anomaly" of S, counted from F_0, linearly increasing with time
- ω' angular distance GF_0, tabulated as function of κ (fol. 382v; *cf.* fig. 8)
- $a = \bar{a}' + \omega'$ "true anomaly," counted from G
- λ "true longitude" of S, counted from ♈0°
- $\delta = \lambda - \bar{\lambda}'$.

On the basis of this arrangement we can now explain

the operation of the tables. Our goal is to find, for a given moment, the true longitude λ of the moon. We can consider as known the mean longitude $\bar{\lambda}'$, the mean anomaly $\bar{\alpha}'$, and the value of $\kappa = \bar{\Delta} - \delta_0$ for the moment in question, since these values can be found in the usual fashion from the tables for 30 Persian years, single years, months, days, and hours (fol. 379ᵛ to 382ʳ). The subsequent tables (fol. 382ᵛ to 384ʳ) for corrections are designed to find δ and thus finally $\lambda = \bar{\lambda}' + \delta$.

Description of the Tables

"First correction" (πρώτη ὄρθωσις): as function of κ (*cf.* the graph fig. 8), i.e., the distance from G to F_0 (fig. 7). This corresponds to the ἐκκέντρου προσθαφαι-

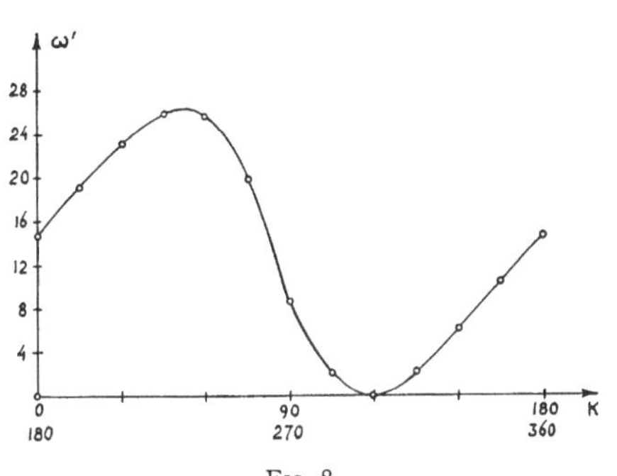

FIG. 8

ρέσεις ἀπογείου of Almagest V,8 col. 3 except for the addition of $\omega_0 = 14;37$ which makes ω' non-negative. The elongation κ is defined as $\bar{\lambda}' - \bar{\lambda}_\odot = \bar{\lambda}_☾ - \bar{\lambda}_\odot - \delta_0$. Since $\delta_0 = 5°$ we have $\kappa = \Delta - 5°$ where $\bar{\Delta}$ is the elongation in the sense of the Almagest. Using the values c_3 of Almagest V,8 column 3 and computing $c_3 + 13;9$ as function of Δ one finds the values of $\omega' \pm 0;1°$ where $13;9 = \max. c_3$.

"Second correction" (δευτέρα ὄρθωσις): ω'' as function of α (*cf.* fig. 9*a*). This function is constructed as follows. We assume the epicycle to be located at the apogee of the eccenter (C_1 of fig. 7) and tabulate the corresponding equation δ_1 as function of α (corresponding to Almagest V,8 col. 4: πλάτους καὶ μήκους προσθαφαιρέσεις ἐπικύκλου) but we add to the result the maximum equation δ_0, thus obtaining a non-negative function. For anomalies α which satisfy $180 \leq \alpha \leq 360$ this is the correction ω''. For $0 \leq \alpha \leq 180$, however, we assume the epicycle to be in the perigee of the eccenter (C_3 of fig. 7) and tabulate $\delta_3 + \delta_0$ as ω''.

"Increment" (πλέον καὶ ἔλαττον): η as function of α (fig. 9*b*). This is the amount of the excess of the epicyclic equation δ_3 in the perigee of the eccenter over δ_1 in the apogee (corresponding to Almagest V,8 col. 5: ἐπικύκλου διαφορά).

"Coefficients of interpolation" (λεπτὰ τοῦ μερισμοῦ); c_1 (πρῶτα) and c_2 (δεύτερα) as function of κ (fig. 9*c*), corresponding to Almagest V,8 col. 6 (διαφορὰ ἑξηκοστῶν), c_1 for $0 \leq \alpha \leq 180$, c_2 for $180 \leq \alpha \leq 360$. This is expressed for c_1 by τοῦτο διὰ μήκους ἤτοι τοῦ κέντρου λαμβάνεται, ὅταν ἄνω τοῦ πλείονος καὶ ἐλάττονος εὑρεθῶσι τὰ ζώδια τῆς ἰδίας [8] "this is taken as a function of elongation or center for the zodiacal signs (0 to 5) which are found (written) above (the table of) the increment" and similarly for c_2 and the signs (6 to 11) written below the table.

Use of the Tables

For a given mean anomaly $\bar{\alpha}'$ (ἰδία κίνησις) and elongation κ we find the true (or "adjusted") anomaly α (διακεκριμένη ἰδία κίνησις) from

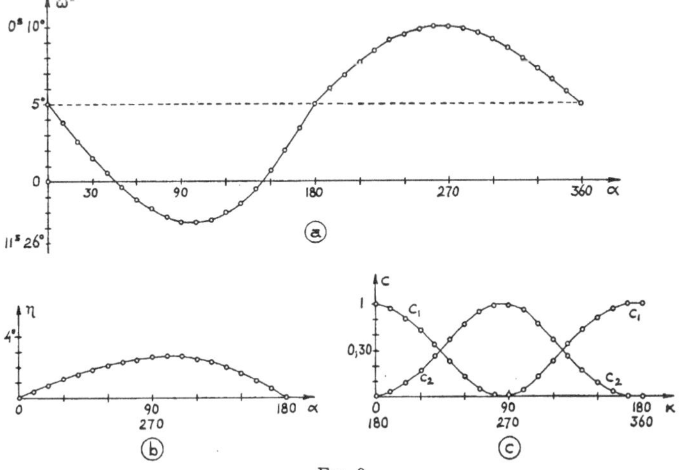

FIG. 9

(1) $$a = \bar{a}' + \omega'(\kappa).$$

With this value of a as argument, we find the second correction $\omega''(a)$ and the increment $\eta(a)$. Furthermore with the given κ

(2) $$c(\kappa) = \begin{cases} c_1(\kappa) & \text{if} \quad 0 \leqq a \leqq 180 \\ c_2(\kappa) & \text{if} \quad 180 \leqq a \leqq 360. \end{cases}$$

Then

(3) $$\delta = \omega''(a) + c(\kappa) \cdot \eta(a)$$

is the final equation and thus

(4) $$\lambda = \bar{\lambda}' + \delta$$

the true longitude.

By means of fig. 7 it is easy to find the justification for this procedure. According to (1) the angle a is the true anomaly counted from the true apogee G of the epicycle, i.e., the apogee as seen from O. We now consider (2) for the first case ($0 \leqq a \leqq 180$). For $\kappa = 355$, i.e. for the apogee of the eccenter, we have $c_1 = 1$ and thus $\delta = \omega'' + \eta$. In this case ω'' is the equation for the perigee of the eccenter, while η represents the increment of the equation if we go from the apogee to the perigee. By adding η in its full amount to ω'' we obtain the equation for the apogee, as it should be for $\kappa = -\delta_0$. For $\kappa = 90 - \delta_0$, i.e. in the perigee of the eccenter, we obtain $\delta = \omega''$ in agreement with the definition of ω''.

For (2) in the second case ($180 \leqq a \leqq 360$) the opposite situation holds. Now ω'' is defined as the equation of the epicycle at the apogee. Indeed, for $\kappa = 355$, we have $c_2 = 0$ and $\delta = \omega''$. For the perigee $\kappa = 85$, however, we find $c_2 = 1$ and thus $\delta = \omega'' + \eta$, which is the value of the equation for the apogee plus the full amount of the increment, hence the equation for the perigee. Intermediate values of the elongation lead to proper intermediate values of the correction $c \cdot \eta$ as is obvious from the graph in fig. 9*c*.

[1] This was realized already by Bullialdus (Astron. Philol. p. 221); also Delambre, HAMA p. 193.
[2] *Cf.* for this *zīj* Kennedy, Survey, No. 84 and 23.
[3] fol. 384[r] above column title.

APPENDIX 7

The planetary tables (fol. 385[v] to 419[r]) show exactly the same characteristics as the lunar tables, the main principle being reduction of the number of steps to a minimum, and norming in such a fashion that all corrections become non-negative.[1]

Using the notation of fig. 10 (E = equant,[2] S = planet, O = observer as in fig. 5) the true longitude of a planet is given by

$$\lambda = \lambda_A + \kappa + \delta.$$

In order to find κ and δ one determines a "first correction" (πρώτη ὄρθωσις) ω_1 as function of the "mean longitude" $\bar{\kappa}$, which is counted from the apogee A (μέση κίνησις) and with it the "adjusted" (διακεκριμένη) quantities

$$\kappa = \bar{\kappa} + \omega_1(\bar{\kappa})$$
$$a = \bar{a} - \omega_1(\bar{\kappa}).$$

With these values as arguments, one can find

$$\lambda_A + \delta = \omega_2(a) + c(\kappa) \cdot \eta(a)$$

where ω_2 is the "second correction" (δευτέρα ὄρθωσις), η the "increment" (πλέον καὶ ἔλαττον) and c the first or second "coefficient" (λεπτὰ γενικά), depending on κ being $\leqq 180$ or $\geqq 180$.

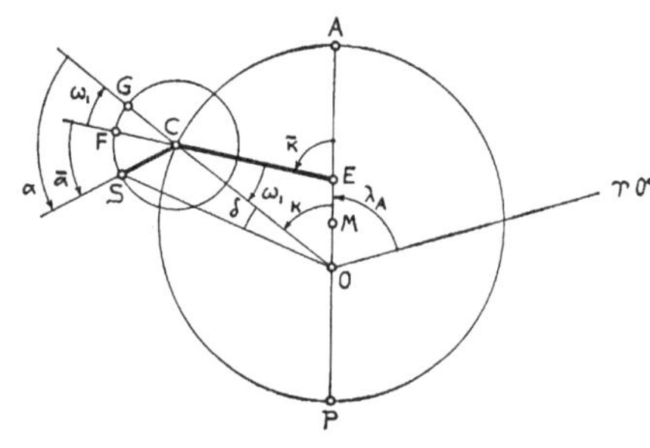

Fig. 10

The tables for the "second correction" ω_2 are constructed as follows. We have to distinguish two cases: (*a*) the epicycle being at the perigee P (μῆκος ἐγγύτερον) and (*b*) at the apogee (μῆκος πορρώτερον). In the Almagest XI,11 we find tabulated the equation δ in the case of mean distance in column (6) (ἀνωμαλίας προσθαφαιρέσεις) from which one obtains the values for the perigee by adding column (7) (διάφορα προσθέσεως), and the values for the apogee by subtracting column (5) (διάφορα ἀφαιρέσεως). From these columns ω_2 is derived in the following way:

for	ω_2 at P	ω_2 at A	$\eta(a)$
$0 \leqq a \leqq 180$	$\lambda_A + (6)$	$\lambda_A + (6) + (5)$	(5)
$180 \leqq a \leqq 360$	$\lambda_A + (6) + (7)$	$\lambda_A + (6)$	(7)

This definition of ω_2 corresponds exactly to the definition of ω'' in the lunar tables from two functions, one for minimum distance (quadratures), one for maximum distance (syzygies). The combined function represents always the smallest possible equation and consequently all subsequent corrections are positive.

[1] All tables can be derived numerically from the tables in the Almagest XI,11 with the exception of ω_1 for Venus and columns (5) and (7) for Mercury. The coefficients of interpolation (λεπτὰ γενικά) are here much less accurate than in the Almagest, column (8).
[2] For Mercury E is rotating (*cf.* Almagest IX,6).

APPENDIX 8

According to the Almagest the latitudes of Venus and Mercury can be found by the addition of three components:

$$\beta = \beta_1 + \beta_2 + \beta_3.$$

These components have to be computed as functions of anomaly a and mean longitude $\bar{\lambda}$ in the following fashion. The tables in XIII,5 give two functions of a:

ἔγκλισις: $b_1(a)$

λόξωσις: $b_2(a)$

and a sinusoidal function of its argument

ἑξηκοστά: $c(\theta)$

where $\theta = \bar{\lambda}$ or $\bar{\lambda} \pm 90$ or $\bar{\lambda} + 180$ in order to account for different combinations of northern and southern components. Disregarding these case distinctions for signs, we can formulate the rules given in Almagest XIII,6 as follows

$$\beta_1 = b_1 \cdot c$$

$$\beta_2 = \begin{cases} b_2 \cdot c & \text{for Venus} \\ (b_2 \pm \tfrac{1}{10} b_2)c & \text{for Mercury} \end{cases}$$

$$\beta_3 = \begin{Bmatrix} 0;10 \\ -0;45 \end{Bmatrix} \cdot c^2 \quad \begin{matrix} \text{for Venus} \\ \text{for Mercury.} \end{matrix}$$

In our manuscript we have two sets of tables which concern planetary latitudes: one from the *Sanjarī zīj*,[1] another from an anonymous work [2] which is probably to be identified with the *'Alā'ī zīj*. The first set follows the procedure of the Almagest, except for some minor modifications. Also the second set is derived from the numerical material of the tables in the Almagest XII,5 but with a slightly different combination of the single terms.

The latitude columns in the *Sanjarī* tables are numbered from 9 to 13. Their headings are:

9: λεπτὰ γενικὰ πλάτους ᾱ: $c_9(\bar{\lambda}) = \sin(90 + \bar{\lambda}) = c(\theta)$ of Almag.

10: πλάτος ᾱ: $b_1(a)$ as in Almagest (ἔγκλισις)

11: λεπτὰ γενικὰ πλάτους β̄: $c_{11}(\bar{\lambda}) = \sin \bar{\lambda}$

12: πλάτος β̄: $b_2(a)$ as in Almagest (λόξωσις)

13: λεπτὰ πλάτους γ̄: $c_{13} = \begin{Bmatrix} 0;10 \\ 0;45 \end{Bmatrix} \cdot c_9{}^2$ for $\begin{Bmatrix} \text{Venus} \\ \text{Mercury.} \end{Bmatrix}$

These numbers have to be combined—with proper signs—exactly as in the Almagest, resulting in

$$\beta = b_1 \cdot c_9 + b_2 \cdot c_{11} + c_{13} \qquad \text{for Venus}$$

$$\beta = b_1 \cdot c_9 + (b_2 \pm 0;6 \cdot b_2) \cdot c_{11} - c_{13} \quad \text{for Mercury.}$$

The question whether the single components should be counted as northerly or southerly is regulated by a system of "signs" (σημεῖα) which are combined algebraically exactly as our + and — signs. Both to c_9 and b_1 as well as to c_{11} and b_2 are assigned signs a or β depending on the semicircle of the anomaly or on the latitude of the center of the epicycle. A product of two equal letters is counted as northerly or positive, while the product of two different letters indicates a southerly latitude or a negative contribution. This procedure constitutes a great simplification over the verbal rules for all combinations of cases in the Almagest.

The second set of tables contains a tabulation of the following functions:

πρῶτον πλάτος: λεπτὰ γενικά : $c(\bar{\lambda})$

πλάτος βόρειον / πλάτος νότιον : $b_1(a)$ as in Almagest (ἔγκλισις)

δεύτερον πλάτος: λεπτὰ γενικά : $c'(\bar{\lambda})$

πλάτος βόρειον / πλάτος νότιον : $b'_2 = b_2(a) + \begin{cases} 0;10 \text{ for Venus} \\ \pm 0;6 \cdot b_2(a) - 0;45 \text{ for Mercury} \end{cases}$

ὄρθωσις πλάτους : ω'.

From these components, with proper signs, the final latitude is computed:

$$\beta = b_1 \cdot c + b'_2 \cdot c' + \omega'.$$

Comparison with the relations resulting from the rules of the Almagest show that one should expect

$$\omega' = \begin{Bmatrix} 0;10 \\ -0;45 \end{Bmatrix} c(c-1) \quad \begin{matrix} \text{for Venus} \\ \text{for Mercury} \end{matrix}$$

but the numerical agreement is bad, particularly since the period is one half of the period of c. I do not understand the reason for these discrepancies.

[1] fol. 352v to 355r for Venus, with explanatory text from Book VIII ch. 3 (fol. 297r,15 to 297v,19) and fol. 355v to 358r for Mercury, with explanatory text fol. 297v,19 to 298v,4.

[2] fol. 412v to 414r for Venus, fol. 420v to 422r for Mercury, and fol. 422v left. Example for Venus: fol. 263r,11 to 263v,21.

APPENDIX 9

Example for the computation of a solar eclipse:[1] A.H. 695 XII 28, Sunday, (= A.D. 1296 Oct. 28) at Daras.[2] From the *Īlkhanī* tables one finds for this day at noontime:

$$\lambda_{\odot} = 7^s\,12;57^\circ \qquad v_{\odot} = 1;1^{\circ/d}$$
$$\lambda_{☾} = 7^s\,14;31 \qquad v_{☾} = 14;5^{\circ/d}$$

consequently the differences

$$\Delta\lambda = 1;34^\circ \qquad \Delta v = 13;4^{\circ/d}.$$

The true conjunction therefore occurred

$$\mu = \frac{24\Delta\lambda}{\Delta v} = \frac{37;36}{13;4} = 2;53^h$$

before noon. This interval is called μῆκος τῆς ὥρας.

For the given solar longitude and $\phi = 38^\circ$ one finds[3] for the length of daylight

$$C = 10;18^h \text{ called } ἡ\ ὥρα\ τῆς\ ἡμέρας\ πάσης$$

and thus for the time from sunrise to noon

$$c = \tfrac{1}{2}C = 5;9^h$$

and for the moment of conjunction $5;9 - 2;53 = 2;16^h$ after sunrise.

The moment of apparent conjunction, however, is influenced by the longitudinal component of parallax. For this correction a table[4] called τῆς ὀρθώσεως τῶν ὡρῶν τῆς ὄψεως has to be used. The horizontal argument represents integer hours from 0^h to 5^h; the vertical argument, minutes from 3 to 60. The tabulated numbers are hours, minutes, and seconds, representing the moment of the apparent conjunction as function of the moment of the true conjunction,[5] both counted from noon. In the present case one finds for a true conjunction which occurs $\mu = 2;53^h$ before noon that the apparent conjunction falls $\nu = 4;21^h$ before noon.[6] Thus the middle of the eclipse is seen at $c - \nu = 0;48^h$ after sunrise.

In order to determine the duration of the eclipse, we have to find the latitude of the moon at conjunction. Using again the *Īlkhanī* tables we find for the date in question and $2;53^h$ before noon (i.e. at true conjunction) the position of the ascending node: $1^s\,15;13^\circ$. Assuming a solar motion of about $1^{\circ/d}$ and an increase of elongation of $\frac{1}{12}\Delta\lambda = 0;5 \cdot 1;34 = 0;7,50^\circ$ between true conjunction and noon. Thus the solar longitude at true conjunction was $7^s\,12;57 - 0;7,50 = 7^s\,12;49^\circ$ and thus the argument of latitude

$$\omega = 7^s\,12;49 - 1^s\,15;13 = 5^s\,27;36.$$

With ω as argument we find in the table τῆς ὀρθώσεως τῆς σελήνης[7] in column η the true latitude (πλάτος) $\beta = +0;12^\circ$.

Again we have to find a correction for parallax. The author says[8] that he has prepared a "ὡριαῖον κανόνιον" in which one enters with the argument ν (here $\nu = 4;21^h$) and which gives a latitudinal parallax β' such that $\beta' - \beta = \beta''$ is the apparent latitude at the apparent conjunction (πλάτος στερεόν). In our case one finds[9] $\beta' = +0;26$ thus $\beta'' = 0;26 - 0;12 = 0;14^\circ$. *Cf.* appendix 10.

With this "final latitude" $\beta'' = 0;14^\circ$ as vertical argument one enters a table for the duration of solar eclipses as prepared "by Samps or indeed by any other competent scholar,"[10] using as horizontal argument the lunar velocity, here $v_{☾} = 14^{\circ/d}$. The result is the half duration (ὥρα πεσοῦσα), i.e., the time from first contact to eclipse-middle, here $0;54^h$.

The solar eclipse under discussion here is No. 5972 of Oppolzer's Canon. According to it the path of totality crosses Mesopotamia, beginning very close to the area of Daras.[11] *Cf.* also Schroeter, Sonnenfinsternisse, map 86*a*.

A similar computation for the lunar eclipse of A.H. 694 VII 13 = A.D. 1295 May 30 (Oppolzer No. 3875) is given fol. 268^v,17 to 269^r,31. The terminology is essentially the same as for solar eclipses.

[1] fol. 270^r,5 to 270^v,27 and fol. 271^v,7 to 272^v,4.

[2] fol. 268^v,19: εἰς πόλιν δάρας ἤτοι τὸ νῦν λεγόμενον ταυρές. Honigmann is wrong in declaring (Ostgrenze, p. 217) that Daras is not Dara in Mesopotamia but a place in the Crimea, since the astronomical data fit only Dara and exclude the Crimea. This is confirmed by fol. 271^v, where $\phi = 38^\circ$ is quoted for the latitude of ταυρές.

[3] From the oblique ascensions for $\phi = 36$ (fol. 436^v) and $\phi = 41$ (fol. 437^r) by linear interpolation.

[4] fol. 425^r.

[5] Parallax at conjunction is a function of three variables: geographical latitude, solar longitude, and hour. Even if the present table were meant for a specific latitude (although not mentioned in the table) there is no entry for the solar longitude. Thus we are dealing with some compromise situation.

[6] Since the elongation increases by $0;32,40^{\circ/h}$ and since the time between true and apparent conjunction amounts to $1;28^h$, the longitudinal component of the parallax must amount to almost $0;48^\circ$.

[7] fol. 343^r.

[8] fol. 270^v,5 f.

[9] The table is given on fol. 271^v for $\phi = 38$.

[10] fol. 270^v,14 f.: τὸ παρὰ σὰμψ ποιηθὲν εἴπερ τινὸς ἄλλου σοφωτάτου ἀνδρός. A table of this type is found on fol. 428^v.

[11] Honigmann, Ostgr. p. 16. The latitude of Daras (= Dārā, near Mardin) is, according to the Russian World Atlas, only about 37;30, not 38.

APPENDIX 10

In the computation of a solar eclipse enters a correction for the latitudinal parallax at the moment of the apparent conjunction. This problem is discussed fol. 271^v,7 to 272^v,4 in connection with the same solar eclipse, total at Daras, (A.D. 1296 Oct. 28) which is the subject of our Appendix 9. The results are tabulated on fol. 271^v (*cf.* table 3).

For the eclipse in question it was found (*cf.* appendix 9):

$$\text{apparent conjunction at } \lambda = 7^s\,13^\circ = ♏13^\circ$$
$$\text{time } \nu = 4;20^h \text{ before noon}$$
$$\text{lunar velocity } v_☾ = 14;5^{\circ/d}$$
$$\text{geogr. latit. } \phi = 38^\circ.$$

Since Theon's tables [1] of parallaxes are computed for the seven climata only, linear interpolation is used for $\phi = 38$ between clima 4 ($\phi = 36$) and clima 5 ($\phi = 41$).[2] The values for the latitudinal parallaxes are found in the tables of fol. 427ᵛ, 428ʳ (also fol. 361 beside longitudinal parallaxes) with references to the beginning of the signs and the epicyclic apogee at syzygies (fol. 361: ἡ ☾ εἰς τὸ μήκιστον μῆκος). This gives the columns 2 to 5 of our table; columns 6 and 7 are the result of interpolation with respect to ϕ, column 8 with respect to λ. Numbers in [ ] are not given in the text.

TABLE 3

①	②	③	④	⑤	⑥	⑦	⑧	⑨	⑩
ὧραι = hours (before noon)	πλάτος = geogr. lat.				πλάτος ταυρές = φ of Daras		πλάτος ζῳδίου σύνοδου		ὡραῖον κανόνιον (latitud. parallax)
	φ = 36°		φ = 41°		φ = 38°		φ = 38		
	λ = 7ˢ	λ = 8ˢ	λ = 7ˢ	λ = 8ˢ	λ = 7ˢ	λ = 8ˢ	λ = 7ˢ 13°		
6ʰ	[0;]14°		[0;]18°		[0;]15[,36]°			6ʰ	
5	15	[0;]23	19	[0;]27	16[,36]	[0;]24[,36]°	[0;]20[,19,36]°	5	[0;]23[20]°
4	18	27	23	30	20[,10]	28[,36]	24[≈23,44]	4	28
3	22	31	26	34	23[,36]	3[2,12]	27[,19,36]	3	31[,30]
2	27	35	30	38	28[,12]	36[,12]	32[≈31,40]	2	37[20]
1	31	38	34	41	32[,12]	39[,12]	36[≈35,14]	1	42
0	35	42	38	43	36[,12]	42[,24]	39[≈38,53]	0	45[30]

Column 8 contains, as is seen from the preceding analysis, the latitudinal parallaxes for $\phi = 38$ and $\lambda = ♏13$, for the hours before noon and the moon at apogee. The hours are once more listed in column 9. We now have to find the latitudinal parallaxes for a lunar position in which the daily velocity is $14;5^{\circ/d}$. To this end one has to use a table which is given on fol. 358ᵛ. Its independent argument (cols. 1 and 2) is the lunar (or solar) anomaly. Column 7 gives the corresponding daily lunar velocity. Thus we can use this column as a column of entry if the lunar velocity $v_☾$ is given, as is here the case. Column 12 is called πλέον ἔλαττον ἰδίου (or λεπτὰ ἀνώμαλα on fol. 428ᵛ), i.e. parallax as function of anomaly α. The values tabulated are coefficients c which increase from $c = 1$ (for $\alpha = 0$) to $c = 1;12$ (for $\alpha = 180$). In our case we are given $v_☾ = 14;5^{\circ/d}$. We find in the table only the values 14;3 (for $\alpha = 4^s\,12^\circ$) and 14;8 (for $\alpha = 4^s\,18^\circ$) but in both cases c has the value 1;10. This means that the lunar parallax for $v_☾ = 14;5^{\circ/d}$ is 1;10 times the parallax found in the tables for the moon in apogee. In this way we can now compute a column 10 in table 3 by multiplying the values in column 8 with the fixed factor $c = 1;10$.

Thus we now know the latitudinal parallax for $\phi = 38$, $\lambda = ♏13$, $v_☾ = 14;5^{\circ/d}$, for integer hours before noon. A simple interpolation furnishes the value for the apparent conjunction, $4;20^h$ before noon. The result $\beta' = 0;26^\circ$ then gives the apparent latitude $\beta'' = \beta' - \beta$ at the middle of the eclipse (*cf.* appendix 9).

[1] fol. 361ʳ: ἔργον τοῦ θαβανῆ. The values agree with the tables published by Halma.

[2] Actually these latitudes are only rounded values; fol. 361ʳ and 427ᵛ give for clima 4: $\phi = 36;22$ (Almagest: 36;0), fol. 428ʳ for clima 5: $\phi = 41$, fol. 361ᵛ: 41;14, Almagest: 40;56.

APPENDIX 11

Book IV, Chapter 1 of the *Sanjarī zīj* [1] gives the rules for the rising azimuth (πλάτος τῆς ἀνατολῆς) of the sun or stars, for given declination δ and at given locality of geographical latitude ϕ. If $\bar{\phi}$ is the complement of latitude (πλάτος τέλειον) then the following general rule for the computation of the rising azimuth α is given:

$$\operatorname{Sin}\alpha = R\,\frac{\operatorname{Sin}\delta}{\operatorname{Sin}\bar{\phi}}.$$

Fig. 11 explains this formula. The upper half of this figure represents a meridian section of the celestial sphere, the lower half a quadrant of the eastern horizon.

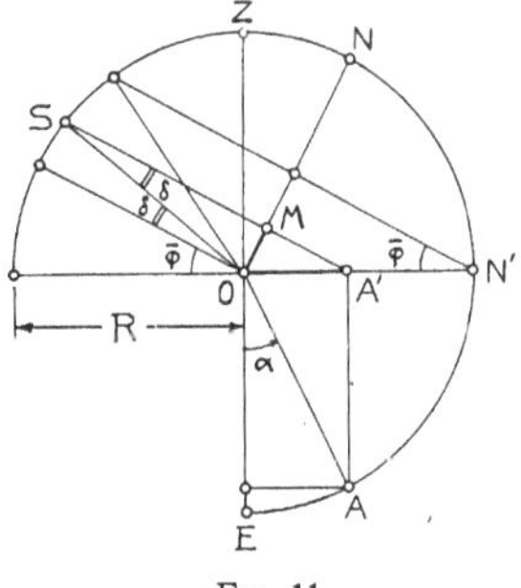

FIG. 11

Since $OA' = \operatorname{Sin}\alpha = R\sin\alpha$ for $ON' = R$, we have for a star culminating at S at a declination [2] δ

$$(1)\quad \frac{OA'}{ON'} = \frac{\operatorname{Sin}\alpha}{R} = \frac{\sin\delta}{\sin\bar{\phi}} = \frac{\operatorname{Sin}\delta}{\operatorname{Sin}\bar{\phi}} \quad \text{or} \quad \sin\alpha = \frac{\sin\delta}{\cos\phi}$$

q. e. d.

It also follows from fig. 11 that stars with $\delta > \bar{\phi}$ are always visible, but never visible if $\delta < -\bar{\phi}$ and furthermore that $\alpha = 90$ for $\delta = \phi$ which is the limit for circumpolar stars. All these cases are mentioned in the text.

[1] fol. 285ʳ,1 to 18.

[2] μετάκλισις in the case of the sun, διάστασις ἀπὸ τοῦ τελείου κύκλου τῆς ἡμέρας or μῆκος in the case of a star.

APPENDIX 12

Let γ_0 be the difference between half the longest daylight, measured in degrees, and 90° at a locality of geographical latitude ϕ. This quantity, called ὄρθωσις τῶν ἡμερῶν, is tabulated fol. 366v and again fol. 373v as function of ϕ from $\phi = 0°$ to $\phi = 60°$. Since γ_0 refers to longest daylight, the solar longitude is assumed to be ♋ 0°. For other longitudes the difference γ between half the length of daylight and 90°, i.e. the "ascensional difference," is found by means of a coefficient[1] c (γενικὰ λεπτά) such that

$$\operatorname{Sin} \gamma(\lambda, \phi) = c(\lambda) \operatorname{Sin} \gamma_0(\phi).$$

The function Sin γ (τραχηλαία τῆς ὀρθώσεως τῶν ἡμερῶν) can also be found[2] directly from

$$(2) \qquad \operatorname{Sin} \gamma = R \frac{\operatorname{Sin} \phi \operatorname{Sin} \delta}{\operatorname{Sin} \bar{\phi} \operatorname{Sin} \bar{\delta}} = R \tan \phi \tan \delta.$$

For $\delta = \epsilon$ one obtains Sin γ_0 which is also tabulated.

The correctness of (2) can be seen from fig. 11, which shows that

$$\mathrm{A'M} = \mathrm{MS} \sin \gamma = \mathrm{OA'} \cos \bar{\phi} = R \sin a \sin \phi.$$

Since $\mathrm{MS} = R \cos \delta$ we have with (1)

$$\sin \gamma = \frac{1}{R \cos \delta} R \frac{\sin \delta}{\cos \phi} \sin \phi = \tan \phi \tan \delta$$

q. e. d.

[1] Tabulated fol. 366v and 373v.
[2] fol. 285r,19 to 285v,3.

APPENDIX 13

From Book V Chapter 2 of the *Sanjarī zīj*[1] we learn the exact definition of the term "distance" (μῆκος or διάστασις) of a star from the equator. One has accordingly to distinguish two cases: (*a*) The star lies in the ecliptic, at a longitude λ. Then the "distance" is the same as the "first declination" (μετάκλισις πρώτη) of the point λ, i.e., in modern terms, the declination of the star. (*b*) The latitude β of the star is not zero. Then the "distance" is given by $\beta + \delta_2$ where δ_2 is the "second declination" (μετάκλισις δευτέρα; *cf.* fig. 2 p. 12). In other words, for non-ecliptic stars the "distance" from the equator is not measured on a circle of declination but on a circle of latitude.

It is of importance for the problem of identification of stars to realize this dichotomy in definition, as compared with the modern "declination." For the corresponding Islamic terminology *cf.* Nallino, Batt. II p. 324.

[1] fol. 286v,10 to 26.

APPENDIX 14

Fig. 12 represents in its upper half a meridian section of the celestial sphere, in its lower half the horizon, E being the east point. Let S be a point of altitude a, S′ its projection on the horizon, S″ on the meridian plane. Then fig. 12 shows that

$$\mathrm{BA''} = R \sin a \tan \phi.$$

This distance is called σημεῖον τῆς μοίρας τῆς ἀναβάσεως.[1]

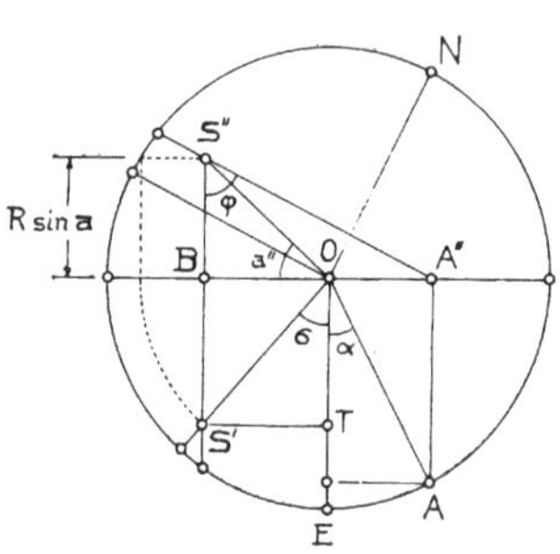

FIG. 12

Consequently,

$$\mathrm{BO} = \mathrm{S'T} = \mathrm{BA''} - \mathrm{OA''}$$

where $\mathrm{OA''} = R \sin a$, a being the πλάτος τῆς ἀνατολῆς (*cf.* appendix 11). BO is the ὄρθωσις τοῦ σημείου.[2] Finally

$$\sin \sigma = \frac{\mathrm{S'T}}{\mathrm{S'O}} = \frac{\mathrm{S'T}}{R \cos a} \quad \text{or} \quad \operatorname{Sin} \sigma = R \frac{\mathrm{S'T}}{\operatorname{Sin} \bar{a}}$$

where Sin σ is called τραχηλαία τοῦ σημείου.[3]

[1] fol. 289v,25.
[2] fol. 289v,29.
[3] fol. 290r,1.

APPENDIX 15

The tables for lunar eclipses, fol. 359v, give digits (δάκτυλοι) and half duration of the partial phase (ὥρα πεσοῦσα) and of totality (ὥρα στάσεως) under the assumption that the moon is at the apogee of the epicycle, which, in turn, must be at the apogee of the eccenter for the syzygies. The same table gives to each of these quantities a correction (ὄρθωσις δακτύλων or ὡρῶν respectively). All these quantities are tabulated as function of the latitude of the moon for the moment of opposition.

If the moon at this moment is not at the apogee of the epicycle then the table[1] μεταβάσεως δ′ καὶ ☾ (fol. 358v) gives a column λεπτὰ τοῦ αὐθημερινοῦ as function of the anomaly a. The tabulated numbers $c(a)$ increase from $c(0) = 0$ to $c(180) = 1$. If d_0 is a value found in the eclipse table fol. 359v and d' the corresponding correction, then

$$d = d_0 + c(a) \cdot d'$$

is the value which corresponds to a position of the moon at anomaly *α*.

[1] The same table is found in the Almagest VI,8 under the title *διορθώσεως κανόνιον* (Heib. I, p. 522). The tabulated numbers are called *διαφόρων ἑξηκοστά*, corresponding to the *λεπτὰ αὐθημερινοῦ* in our text.

APPENDIX 16

According to fol. 293^{r},8 the tropical year which underlies the Maliki calendar has the length of 365;14,27,20, 36,47^{d}. The excess over 365^{d} is therefore a fraction of one rotation, namely

$$c = 6{,}0 \cdot 0;14{,}27{,}20{,}36{,}47 = 1{,}26;44{,}3{,}40{,}42^{\circ}.$$

This value, rounded to 1,26;44 or 1,26;44,4 is called (fol. 367^{r}) *περίσσεια τῶν χρόνων* "excess of the years." The mean value c is then modified by a correction c' (*ὄρθωσις τῆς περισσείας τῶν χρόνων*) which depends on the solar anomaly. On fol. 367^{r} this correction is tabulated separately, whereas $c + c'$ is given on fol. 429^{r} as *περίσσεια τῆς περιφορᾶς τοῦ χρόνου τοῦ* ☉ "excess of revolution of the solar year," a terminology which reflects the Arabic *faḍl al-dawr* "excess of revolution." [1]

[1] *Cf.* Kennedy, Survey, p. 147; Nallino, Batt. I, p. 211.

APPENDIX 17

CODEX VATICANUS GRAECUS 1058, FOL. 261 TO 459

GENERAL DESCRIPTION

According to Heiberg, Byz. Anal. p. 170, our manuscript was written in the fifteenth century. Indeed, fol. 6^{r} contains a reference to "the present year 6917" (= A.D. 1408/9). The latest epoch year that occurs in the part of the codex investigated here is the year Yazd. 715 (= A.D. 1346) as epoch for a catalogue of stars.[1] In the planetary tables the years Yazd. 721 (= A.D. 1352) and Yazd. 781 (= A.D. 1412) are specifically noted on the margin.[2]

At least parts of the manuscript were prepared for use at Byzantium, as, e.g., the inclusion of a table of rising times for this locality [3] proves. Finally, however, the manuscript must have come to southern Italy, since we find some marginal notes [4] which give corrections for the geographical longitude of "Calabria." [5]

Beside the table of contents, which had been given by Heiberg (1899) in his Byz. Anal. p. 170-172, a short summary was published by Tannery (1888) in his report on a mission to Italy.[6] A very close parallel to our text is Vat. gr. 211, fol. 122 to 233 which is described in vol. I of the Cod. Vat. Graeci, p. 264-269.

1. A closer inspection shows that the text is of a rather composite character. A first section (fol. 261 to 273^{v}) is obviously related to the *'Alā'ī zīj*,[7] which is quoted on several occasions, though there are also references to the *Sanjarī zīj*,[8] the *Īlkhānī zīj*,[9] and to the methods of Shams ad-Dīn of Buchara.[10] The examples in this part concern the years A.D. 1295/1296, in particular a lunar and a solar eclipse computed for the circumstances at Daras in upper Mesopotamia.[11]

The tables which belong to these examples form the second part of the collection of tables that follows the introductory text, namely the tables from fol. 375^{r} to at least fol. 422^{v}. These tables are normed for the epoch year Yazd. 541 (= A.D. 1172) and for the meridian of 84°. Tables and computations are based on Persian years. It seems plausible to assume that at least the tables come from the *'Alā'ī zīj*.

2. On fol. 273^{v} the introductory text changes abruptly to a summary, or translation, of the *Sanjarī zīj*, extending to fol. 316^{r} and covering all its twelve books, each of which contains several chapters and subsections. Comparison with the summary given by Kennedy in § 12 of his Survey shows that our version is not identical with the Arabic version,[12] in spite of a close parallelism in the major layout.

The tables which belong to this part of our manuscript form the beginning of the set of tables, i.e., fol. 332 probably to fol. 366^{v}. The epoch year for the motions of the celestial bodies is A.H. 421 (= A.D. 1030) and the years are always Arabic lunar years. The epoch of the star catalogue (fol. 363^{v}) is A.H. 509 (= A.D. 1115/6).

Examples for the use of the *Sanjarī zīj* are given on fol. 322^{r} to 325^{v}, concerning again the years A.D. 1295/6 as is the case for the *'Alā'ī zīj* and also the year A.D. 1254 with a reference to Shams ad-Dīn of Buchara. This shows that it is not accidental that text and table of the *'Alā'ī zīj* are combined in the same manuscript with text and tables of the *Sanjarī zīj*. Shams ad-Dīn, who is once [13] called "our teacher" apparently transmitted both methods to the author or scribe of the present work.

3. Between the summary of the *Sanjarī zīj* and the examples are inserted a few pages (fol. 316 to 321) which have nothing to do with the astronomical tables. They form a primitive treatise on elementary astronomical concepts, with a philosophical introduction on matter and motion on Aristotelian lines. This whole section is uninfluenced by Islamic astronomy.

4. Following the examples which belong to the *Sanjarī zīj* are pages which hardly belong to the major section, although they are related to Islamic astronomy (fol. 326 to 331). The main subject concerns the arrangement and computation of ephemerides, with examples for the year A.H. 701 (= A.D. 1302). A concluding paragraph deals with the Uighur calendar for the year A.D. 1290.

5. Besides the two major *zījes* the tables contain material which probably was added more or less accidentally to these works. Astrological material is contained in the tables fol. 367 and 369; the tables of ascensional differences are given in two versions (fol. 366^v and 373^v); a table for the pseudo-comet *al-kaid* (fol. 374) intruded just at the beginning of the *'Alā'ī* tables and at the end we find, after subsidiary astrological and trigonometric tables (fol. 429 to 440), a catalogue of stars for Yazd. 715 (= A.D. 1346), a version of *Abū Ma'shar*'s Paranatellonta (fol. 454 to 457), followed by a list of "famous cities" (fol. 458 and 459).

[1] fol. 441^r. A star catalogue with exactly the same date is probably the work of Eleutherius Eleus; *cf.* CCAG 5,1, p. 36 note 1, p. 54 fol. 266^v; CCAG 6, p. 2 fol. 30, p. 51, fol. 149^v.

[2] fol. 379^v, 384^v, 392^v.

[3] fol. 438^v.

[4] fol. 378^r, 382^r, 416^r.

[5] *Cf.* the index of geographical references (p. 39). For the important connection between Byzantium and Calabria *see* Setton, Byz. Backgr.

[6] Mém. Sci. 2, p. 316 f.

[7] Kennedy, Survey, No. 42; the accurate date of this work is unknown.

[8] Kennedy, Survey, No. 27 and summary § 12; the author is al-Khāzinī (about A.D. 1120).

[9] Kennedy, Survey, No. 6 and summary § 13; the author is Naṣīr ad-Dīn aṭ-Ṭūṣī (about A.D. 1270).

[10] As Suter (MAA p. 161, No. 397, and p. 219 note 80) suggested, this is probably the same person as Shams ad-Dīn Mīrak al-Bukharī; *cf.* also Kennedy, Survey, p. 129, No. 25.

[11] Three examples (fol. 265^v, 266^r) for the year A.D. 1341 are suspect, since the given lunar positions do not agree with this date.

[12] Cod. Vat. Arab. 761. *Cf. also* Nallino, Batt. I, p. 279 ff.

[13] fol. 268^r,31.

TABLE OF CONTENTS

The numbers refer to the folios of Vat. Gr. 1058; numbers of sections in ( ) are not given in the text

A. *Text*

Folio	Division	Content
278^{v},3		1. Weekday of the first day of years and of months
279^{r},21		2. Arabic years and Roman, Persian, Maliki years
280^{r},9	Chapter 5.	Easter and other Feasts
280^{r},16		1. The lunar mansions
280^{v},1		2. Christian feast days
280^{v},12		3. Christian feast days, contin.
280^{v},27		4. Islamic Easter and other feast days
281^{v},15		5. Persian Easter and other feast days
282^{r},5		6. The Persian names of the days
282^{r},13		7. (Syrian) Christian feast days
282^{v},1	*Book II.*	Trigonometric Functions
282^{v},4	Section 1.	Rules for linear interpolation (1.)
282^{v},28		(2.) continued
283^{r},10	Section 2.	(1.) Definition for Sin θ and sagitta
283^{v},1		(2.) Arc and Sin θ
283^{v},13		(3.) Arc and sagitta
283^{v},29	Section 3.	The shadow function $(12 \cdot \cot \theta)$
283^{v},32	*Book III.*	Table of Contents
284^{r},2	Chapter 1.	First and second declination
284^{r},14	Chapter 2.	Determination of the geographical latitude (1.) from culmination of sun or star
284^{r},22		(2.) from always visible stars or from sun at solstices or equinoxes
284^{v},1	Chapter 3.	Altitude of sun and of stars
284^{v},11	Chapter 4.	Rising time for sphaera recta
284^{v},29	*Book IV.*	Table of Contents
285^{r},1	Chapter 1.	Azimuth of rising point of sun or stars
285^{r},19	Chapter 2.	(1.) Length of daylight from ascensional difference γ
285^{v},4		Example for $\phi = 45$, $\lambda =$ ♌ $0°$ ("Scholion of Branā")
285^{v},18		(2.) sagitta of γ
285^{v},23	Chapter 3.	(1.) Length of daylight and night
286^{r},3		(2.) Equinoctial and seasonal hours
286^{r},13	Chapter 4.	Oblique ascension
286^{r},19	*Book V.*	Table of Contents
286^{r},27	Chapter 1.	Longitude of fixed stars, precession
286^{v},4		Alternative procedure by Shams (of Buchara)
286^{v},10	Chapter 2.	Distance of fixed stars from equator (1.)
286^{v},26		(2.) Alternative procedure
287^{r},6	Chapter 3.	Simultaneously culminating degree
287^{r},27	Chapter 4.	Simultaneously rising degree; (1.) rising
287^{v},7		(2.) setting
287^{v},12	Chapter 5.	Time of rising
287^{v},26	*Book VI.*	Table of Contents
287^{v},30	Chapter 1.	(1.) Arc since sunrise from altitude at given moment
288^{r},16		(2.) Altitude at given moment from arc since rising
288^{r},23		(3.) At night
288^{r},28		Procedure by Shams (of Buchara)
288^{v},8		(4.) Seasonal hours since rising
288^{v},22	Chapter 2.	(1.) Hour from rising times
289^{r},3		(2.) Rising times from culminating point
289^{r},8	Chapter 3.	Arc since sunrise or sunset
289^{r},18	Chapter 4.	The 12 loci
289^{v},22	Chapter 5.	(1.) Altitude and circle of declination
289^{v},30		(2.) continued
290^{r},7		(3.) Case of the polar region
290^{r},15	Chapter 6.	Determination of the terrestrial meridian
290^{v},5	Chapter 7.	Determination of the qibla
291^{r}		Table for the qibla for 30 cities
291^{r},12	*Book VII.*	Table of Contents, quoting ʿAbd ar-Raḥmān al-Khāzinī
291^{v},11	Chapter 1.	Computation of mean motion from the tables for the longitude 90° East (1.) Motion in longitude
292^{r},31		(2.) The apogees
292^{v},7	Chapter 2.	Correction for geographical longitudes different from 90°. (1.) In general
292^{v},28		(2.) astrological practice
293^{r},6	Chapter 3.	(1.) Maliki years, Roman and Persian years
293^{r},16		(2.) Days of the week
293^{v},4		(3.) Ordinary and intercalary years
293^{v},10	Chapter 4.	General description of the solar, lunar, and planetary tables
294^{r},25	*Book VIII.*	Table of Contents; general arrangement of the argument of the tables
294^{v},18	Chapter 1.	On the longitude of sun, moon, and planets
294^{v},20		1. Computation of the solar longitude
295^{r},4		2. Computation of the longitude and latitude of the moon
295^{v},13		3. The lunar nodes
295^{v},16		4. Computation of the longitude of the five planets
296^{r},15	Chapter 2.	On direct and retrograde motion. (1.) First and second station
296^{r},27		(2.) Duration of direct motion
296^{v},1		(3.) Duration of retrogradation
296^{v},11	Chapter 3.	On latitudes
296^{v},12		1. Lunar latitude
296^{v},26		2. Latitude of the outer planets
297^{r},15		3. Latitude of Venus
297^{v},19		4. Latitude of Mercury

298^{v},4 Chapter 4. On solar and lunar velocity and diameter (1.) Definitions
298^{v},9 (2.) Solar diameter from velocity
298^{v},13 (3.) Lunar diameter from velocity; shadow
298^{v},20 (4.) The diameters by means of the tables

299^{r},1 *Book IX.* On Parallax
299^{r},3 Chapter 1. 1. Angle between ecliptic and circle of altitude at the ascendant
299^{r},13 2. At general position
299^{r},18 3. On the three cases of angles
299^{v},14 4. Parallax, from the tables
299^{v},27 5. Components of parallax in longitude and latitude
300^{r},4 Chapter 2. Parallax from Theon's tables. (1.) Use of the tables
300^{r},25 (2.) continued
300^{v},7 (3.) ?
300^{v},19 (4.) Interpolation for geographical latitude
301^{r},5 (5.) Correction for lunar anomaly
301^{r},12 Chapter 3. (1.) Direction of the longitudinal parallax
301^{r},17 (2.) Direction of the latitudinal parallax

301^{v},6 *Book X.* On conjunctions and oppositions of sun and moon
301^{v},8 No. 1. (1.) Determination of the hour
302^{r},10 (2.) Variant
302^{r},15 (3.) Determination of the longitude
302^{v},2 No. 2. Lunar eclipses
302^{v},3 Chapter 1. Whether an eclipse will occur or not
302^{v},5 1. Conditions for latitudes
302^{v},12 2. Condition for diameter
302^{v},18 3. Partial eclipses
302^{v},30 4. Duration of the phases of an eclipse
303^{r},11 5. Duration of totality
303^{r},26 Chapter 2. Lunar eclipses from the tables; (1.) Magnitude
303^{v},15 (2.) Time of the eclipse
303^{v},22 No. 3. Solar eclipses
303^{v},23 Chapter 1. Introduction
303^{v},31 1. Interpolation of parallax for geographical latitude
304^{r},mrg.3 2. Interpolation of parallax for longitudes within a zodiacal sign
304^{r},mrg.14 3. Correction for lunar anomaly
304^{r},3 Example of interpolation for parallax at $\phi = 38$ and $\lambda_{\odot} =$ ♌ 25
304^{r},13 Chapter 2. Introduction
1. Methods of computation; correction for parallax
305^{v},6 2. Whether an eclipse will occur or not; magnitude
305^{v},34 (3.) Duration; magnitude, linear and area

306^{r},19 *Book XI.* Visibility of moon and planets; Introduction
306^{v},16 Chapter 1. Visibility of the moon
306^{v},17 1. Position of sun and moon at sunset
307^{r},4 2. Parallax
307^{r},9 3. Equation of time
307^{r},15 4. Correction for latitude
307^{r},33 5. Ripeness of the crescent
307^{v},4 6. Delay of moonset after sunset
307^{v},10 7. Arc of sun below horizon at moonset
307^{v},25 (8.) Altitude of the first crescent
308^{r},4 Chapter 2. First visibility of the moon; four angles
308^{r},31 Chapter 3. First visibility of the moon. (1.)
309^{v},17 (2.) Different criteria
309^{v},29 (3.) continued
309^{r},9 Chapter 4. First visibility of the moon. (1.)
309^{r},17 (2.) By observation with an instrument
309^{r},25 Chapter 5. First and last visibility of the planets. (1.) Introduction; visibility limits according to the Hindus and to Ptolemy
309^{v},8 (2.) Appearance and disappearance from tables
309^{v},23 (3.) Date of appearance and disappearance
310^{r},2 Chapter 6. First visibility of the moon
310^{r},6 1. Criterium for visibility
310^{r},22 2. Variant

310^{v},8 *Book XII.* Introduction on the operation with years for astrological purposes
310^{v},16 Chapter 1. (1.) On rising times and ascensional differences
310^{v},23 (2.) Entry of the sun into a zodiacal sign
311^{r},15 (3.) Excess of revolution
311^{r},30 (4.) Rising times for different localities
311^{v},13 Chapter 2. On the configuration of the celestial bodies
311^{v},18 (1.) Distance from a center
312^{r},1 (2.) Latitude of the circle of motion
312^{r},10 (3.) The aspects
312^{v},11 (4.) Further astrological applications
313^{v},10 Chapter 3. On the motion of hyleg; introduction and table
313^{v},31 1. The arc of motion of the hyleg
314^{r},8 (2.) The degree of the hyleg

315^{v},1 Chapter 4. On the motion of the astrological τύχη
315^{v},4 1. The three types of motion
315^{v},25 2. continued
316^{r},4 3. continued
316^{r},10 4. continued; end of the *Sanjarī zīj*.

316^{r},24 On the shape and matter of the stars
316^{v},8 On motion
316^{v},28 On the circular motion, obliquity of the ecliptic, list of stars in constellations
318^{r},14 The motions of the celestial spheres
318^{v},8 The sun
319^{r},25 The spheres of the moon
319^{v},12 The spheres of the planets from Saturn to Venus
319^{v},28 Mercury
320^{r},6 Stations and retrogradations
320^{r},17 Latitudes
320^{r},29 Parallax
320^{v},14 The lunar phases
320^{v},24 Lunar eclipses
320^{v},30 Solar eclipses
321^{r} (except for two lines) and 321^{v}: blank

322^{r},1 On the determination of the positions of the celestial bodies according to the *Reliable Sanjarī zīj*
322^{r},2 Longitude of the sun
322^{v},6 Example for A.H. 695 VI 10 (= A.D. 1296 Apr. 15)
323^{r},9 Day of the week
323^{v},8 Example for A.H. 694 VII 24 (= A.D. 1295 June 10)
323^{v},28 Longitude of the sun, method for the *Sanjarī zīj* of Shams ad-Dīn, for Buchara, A.H. 652 IV 22 (= A.D. 1254 June 11)
324^{v},12 Position of the moon for A.H. 652[1] IV 22 (A.D. 1254 June 11) for *Sanjarī zīj*
325^{v},1 Position of Saturn for A.H. 652 IV 22 (= A.D. 1254 June 11)
326^{r},15 On the arrangement of tables
326^{r},16 Chapter 1. Classification of zodiacal signs
326^{r},26 Chapter 2. List of abbreviations
326^{r},32 Chapter 3. Arrangement of lines and columns in tables
326^{v},18 Chapter 4. Use of red and black writing
326^{v},32 On the composition of an ephemeris
327^{r},17 Example: ephemeris beginning with A.H. 701 VII 11 (= A.D. 1302 March 12)
327^{r},21 Parameters at epoch
327^{v},9 Ephemeris for the sun
328^{v},16 Ephemeris for the moon
329^{v},5 Ephemeris for Saturn and the other planets
330^{v},11 The ascending lunar node
330^{v},18 Hours and geographical longitude
330^{v},27 Ephemeris for Mars when retrograde
330^{v},29 Aspects
331^{r},6 The beginning of Arabic months. Example: Ramadan of A.H. 701 (= A.D. 1302 Apr. 30)
331^{r},31 On the entry of the moon in a zodiacal sign
331^{v},18 On the years of the Mongols, Chinese, Uighur, with example for the year Alexander (= Sel. era) 1613 (= A.D. 1301/2)

B. *Tables*

332 days in months: Arabic, Persian, Syrian, Maliki; days of week
333^{r} concordance between Arabic, Persian, Roman, and Maliki years
333^{v} entry of the sun into lunar mansions for the Roman year (= Seleucid era) 1386 (= A.D. 1075/6); Christian feasts
334^{r} extremal and mean motions of the seven planets; retrogradations; general rules for the composition of an ephemeris
334^{v},335^{r} epoch values for sun, moon, planets, and precession for 30 Arabic years, beginning with A.H. 421 (= A.D. 1030)
335^{v},336^{r} single years
336^{v},337^{r} months and hours
337^{v},338^{r} single days
338^{v} coordinates of "famous cities"
339^{r} equation of time
339^{v},340^{r} sun, equation of center
340^{v} to 343^{r} moon, equations and latitude
343^{v} to 346^{r} ♄, equations and latitude
346^{v} to 349^{r} ♃, equations and latitude
349^{v} to 352^{r} ♂, equations and latitude
352^{v} to 355^{r} ♀, equations and latitude
355^{v} to 358^{r} ☿, equations and latitude
358^{v} velocities, apparent diameter, for sun and moon
359^{r} planets, phases for clima 4
359^{v} lunar eclipses; magnitudes, duration
360^{r} parallax in altitude, for syzygies only
360^{v} to 361^{v} parallax in longitude and in latitude for climata 3 to 5
362^{r} magnitude of solar eclipses and corrections for variable lunar distances, horizontal parallax (?) for climata 3 to 5
362^{v} lunar visibility as function of anomaly and visibility conditions
363^{r} depression of the sun, altitude of the moon, for first visibility
363^{v} to 364^{v} mean motions of the seven planets in months

of the Maliki calendar; fixed stars for 1427 (Seleucid era) = A.H. 509 (= A.D. 1114/5)

364^v,365^r chronological tables for Maliki years (in 20-year steps) and the Roman, Persian, Arabic calendar; motion of κάϊτ

365^v Sin θ and Cot θ

366^r first and second declination of the sun (*cf.* fol. 430^v)

366^v ascensional differences (*cf.* fol. 373^v)

367^r to 368^r "excess of revolution" and related astrological tables

368^v planetary periods

369 beginning of Arabic years; explanatory text for astrological purposes

370^r to 371^r weekdays for Arabic, Persian, Jewish years

371^v to 373^r chronological tables for the Roman, Arabic, Persian calendar

373^v ascensional differences (*cf.* fol. 366^v)

374 κάϊτ (published: *Jour. Amer. Orient. Soc.* **77**: 213)

375^r to 376^r mean motions and epoch values for the seven celestial bodies, for Yazd. 541 (= A.D. 1172) and for geogr. long. 84 [2]

376^v blank

sun: 377^r epoch values for mean motion and apogee, for 30 Persian years, single years, and hours

377^v,378^r mean motion for months and days, correction for geogr. long.

378^v,379^r anomalistic motion

moon: 379^v epoch values for mean motion of $\bar{\lambda}$, $\bar{a}$, elongation, for 30 Pers. years, single years

380^r to 382^r mean motion for months, days, hours; correct. for geogr. long.

382^v 1st correction

383 2nd correction

384^r increment and coefficients

384^v,385^r ascending node, mean motions; correct. for geogr. long.

Saturn: 385^v epoch values for mean motion of $\bar{\lambda}$, $\bar{a}$ for 30 Persian years, single years, hours

386^r to 387^r mean motion for months and days; correct. for geogr. long.

387^v,388^r 1st correction

388^v for perigee, 2nd correction

389^r for perigee, increment and coefficients

389^v for apogee, 2nd correction

390^r for apogee, increment and coefficients

390^v,391^r 1st and 2nd stations

391^v,392^r latitude

Jupiter: 392^v to 399^r, arrangement as with Saturn

Mars: 399^v to 406^r, arrangement as with Saturn

Venus: 406^v to 412^r, arrangement as with Saturn, except for latitudes:

412^v,413^r 1st latitude

413^v,414^r 2nd latitude

Mercury: 414^v to 422^r, arrangement as with Venus

422^v correction for latitude of ♀ and ☿; diameters of the planets

423^r reciprocals of hourly increase of anomaly for the planets

423^v 24 divided by length of daylight

424^r solar motion for hours and for minutes as function of the daily velocity

424^v lunar motion in latitude for eclipses for 30 Arabic year, single years, and months, beginning with A.H. 301 (= A.D. 913); ecliptic limits

425^r ecliptic limits, correction of the time of conjunction for longitudinal parallax

425^v,426^r solar and lunar diameter; sum of radii of shadow and moon as function of lunar and solar velocity

426^v eclipse magnitudes as function of lunar diameter and latitude

427^r duration of lunar eclipses

427^v,428^r parallax in latitude for climata 2 to 7

428^v eclipse magnitudes

429^r to 430^r "excess of revolution"; ἐναλλαγή

430^v 1st and 2nd declination of the sun (*cf.* fol. 366^v)

431^r to 433^r lunar latitude, Tan δ, noon altitudes of the sun, shadow lengths, Sin θ, sagitta

433^v to 438^v rising times for sphaera recta, 7 climata, and Byzantium

439^r to 440^v sexagesimal multiplication tables

441^r to 453^v catalogue of stars for era Yazd. 715 (= A.D. 1346) giving λ, β, magnitude, and κρᾶσις

454^r to 457^v paranatellonta [from Abū Ma'shar; *cf.* *Acad. royale Belgique, Bull. Cl. des lettres* 5e sér., **43**: 133-140, 1957]

458^r to 459^v "famous cities"

460^r beginning of a treatise on the astrolabe [3]

[1] The text says incorrectly A.H. 695 IV 22 (= A.D. 1293 Apr. 2) = Roman year 6800 (i.e. Byzantine World Era), then A.H. 692, but finally computed for A.H. 652.

[2] The same data apply to all tables from fol. 377^r to 422^v.

[3] By Nicephoras Gregoras (as Tannery recognized); *cf.* A. Delatte, Anecdota Atheniensia II: 195 ff., Paris, 1939.

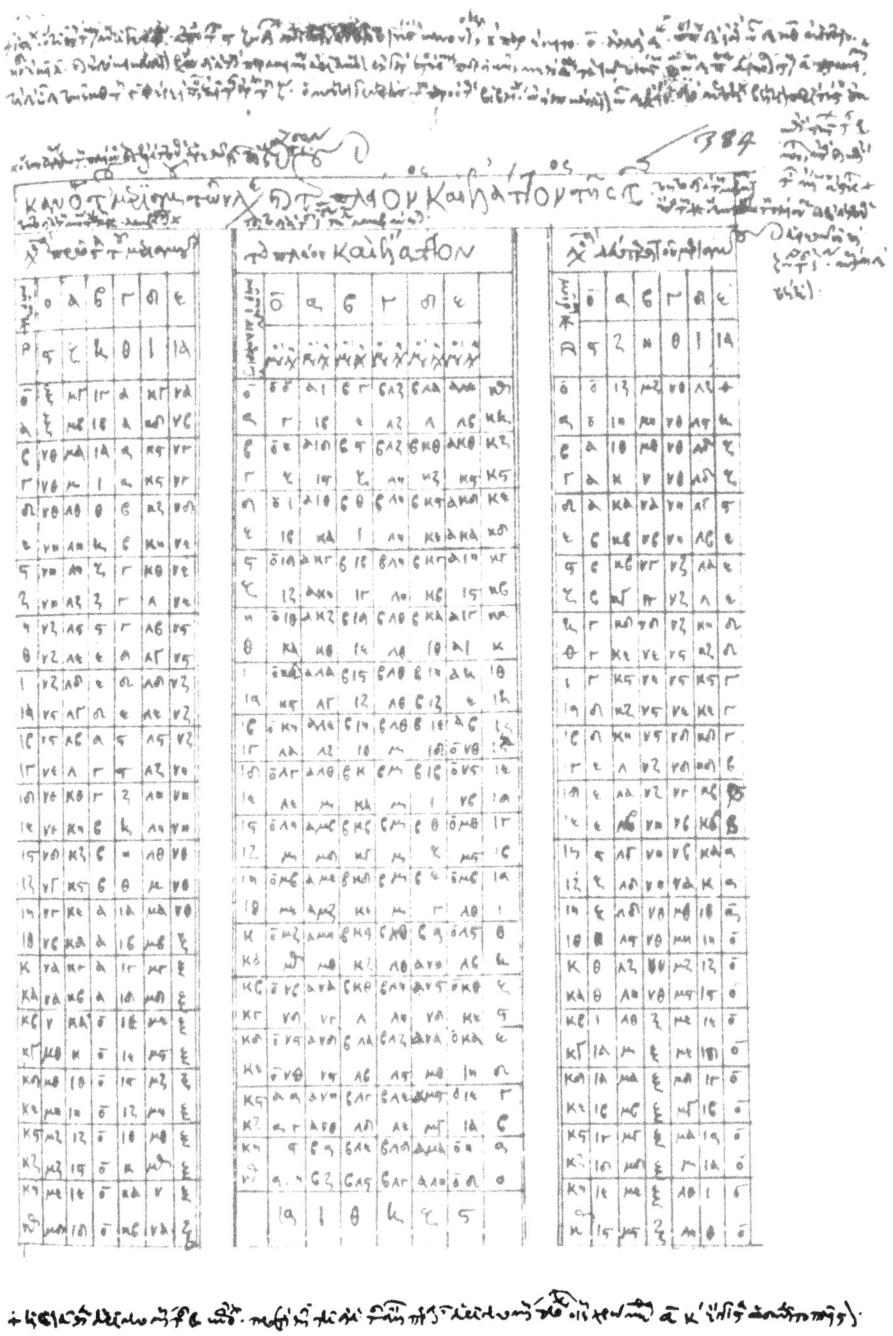

FIG. 13. Vat. gr. 1058, fol. 384r.

DATES QUOTED IN VAT. GR. 1058, FOL. 261 TO 459

The numbers refer to the folios

A. SPECIFIC DATES

A.D.	
1143 April	Roman year (= Seleucid era) 1454 Nisan 22: 280^{r},23 f.
1254 June 11	A.H. 652 IV 22: 323^{v},31/324^{r},1 (longitude of the sun) 325^{v},2 (longitude of Saturn)
1292/3	A.H. 692: 324^{v},14
1295 May 30	A.H. 694 VII 13: 268^{v},18 (lunar eclipse)
1295 June 10	A.H. 694 VII 24: 323^{v},8 f. (weekday)
1295/6	A.H. 695 = Roman year (Byz. World era) 6803: 324^{v},13
1296 Feb. 10	Yazd. 665 II 10: 261^{r},16 (longitude of the sun) 262^{r},1 (longitude of the moon) Roman year (Byz. World era) 6805 Feb. 10 = Yazd. 665 II 10: 263^{r},1 f. (latitude of Jupiter)
1296 Feb. 12	Yazd. 665 II 12: 263^{v},29 f. (station of Saturn)
1296 April 15	A.H. 695 VI 10: 322^{v},6 f. (longitude of the sun)
1296 Oct. 28	A.H. 695 XII 28: 270^{r},6 f.; 271^{v},8 f. (solar eclipse)
1297(?) July 16	Roman year 168(?) (error for 1608 Sel. era?) July 16: 285^{v},4 f. (entry of the sun into Leo)
1301/2	Roman year (= Sel. era) 1613: 331^{v}, 23 Roman years from Alexander: 331^{v},20
1302 March 12	A.H. 701 VII 11: 327^{r},19; 328^{r},20 f. (longitude of the sun) 328^{v},30; 329^{r},20 f. (longitude of the moon) 329^{v},7 f. (longitude of Saturn)
1302 Apr. 30	A.H. 701 Ramadan (IX) 1: 331^{r},8
1341 Aug. 16 and 17	Roman year (= Sel. era) 1652 Aug. 16 and 17: 266^{r},18,21; 266^{v}, 24,25
1341 Aug. 21	Roman year (= Sel. era) 1652 Aug. 21: 266^{r},9
1352	Yazd. 721: 384^{v},marg.; 392^{v},marg.
1376 July 17, 1379 May 16, 1386 Jan. 1	solar eclipses; *cf.* p. 7 s.v. δάκτυλος, note 6
1412	Yazd. 781: 379^{v},marg.

B. EPOCH YEARS IN TABLES

A.D.	
622	A.H. 1: 336^{v}/337^{r} (apogees of the sun and the planets)
632	Yazd. 1: 374^{r}; 377^{r}; 379^{v}; 384^{v}; 385^{v}; 392^{v}; 399^{v}; 406^{v}; 414^{v}
913/4	A.H. 301: 424^{v}
1030	A.H. 421: 334^{v}
1058	A.H. 450 = Yazd. 426 = Roman year (= Sel. era) 1369: 333^{r}
1075/6	Roman year (= Sel. era) 1386: 333^{v},3 Roman year (= Sel. era) 1386 = Yazd. 444 = A.H. 467 = era Melixa 0: 364^{v}
1115	A.H. 509: 286^{r},29; 286^{v},4 A.H. 509 = Roman year (= Sel. era) 1427: 363^{v} (longitudes of fixed stars)
1171/2	Yazd. 540 = Roman year (= Sel. era) 1482 = A.H. 566: 372^{v}
1172 Feb. 1	Yazd. 541 (I 0) Tuesday: 375^{r}; 376^{r} (values of all parameters for the following tables)
1302 March 12	A.H. 701 VII 11 = Adar (VI) 12 = March 12 = Khordād (III) 12: 327^{r},19 f.
1346	Yazd. 715: 441^{r} (catalogue of stars)
1367 Sept. 1	A.M. 6876 (= Byz. World era): p. 7 s.v. δάκτυλος, note 6

GEOGRAPHICAL REFERENCES IN VAT. GR. 1058, FOL. 261 TO 459

The numbers refer to the folios [1]

Buchara *see* μπουχαρâ

βυζάντιον: $\phi = 43$ for the clima διὰ βυζαντίου: 438ᵛ

δάρας: τὸ νῦν λεγόμενον ταυρές: 268ᵛ,19
- ταυρές: 270ʳ,8
- $\phi = 38$: 271ᵛ,11, table; 304ʳ,9 f., marg. 3
- $l = 82$: 272ᵛ,22
- $l = 72$; according to *'Alā'ī* and *Sanjarī zīj*: 272ᵛ, 23
- τῆς πόλεως ἡμῶν τὸ πλάτος ἦν τόσον λη′: 301ᵛ,4

θάλασσα τῆς δύσεως: 323ᵛ,31

καλαβρία: $l = 38$: marg. notes 378ʳ,416ʳ
- $\phi = 38$ (error for 39): 382ʳ,marg.
- $l = 39$ (error for 38): 382ʳ,marg.

κωνσταντινόπολις: 265ᵛ,10
- $l = 49$;50: 261ᵛ,16 f.
- $\phi = 45$: 265ʳ,17,27 f.
- [$l = 56$]: marg. notes 378ʳ, 382ʳ

μακκâ: 290ᵛ,9,13,16 etc.

μπουχαρâ: $l = 87$, $\phi = 39$: 323ᵛ,30 f.

ταυρές *see* δάρας

Latitude: $\phi = 33$: τοῦ μέσου τῆς οἰκουμένης: 311ᵛ,11,13
- $\phi = 37$ (error for 38): 307ᵛ,18
- $\phi = 38$ *see* δάρας

Longitude: $l = 52$: marg. notes 378ʳ, 382ʳ
- $l = 84$: for *'Alā'ī zīj*: 261ʳ,14; 261ᵛ,13 f. for epoch values of tables: 375ʳ; 376ʳ; 379ᵛ; 384ᵛ; 385ᵛ; 392ᵛ; 399ᵛ; 4C5ᵛ; 414ᵛ
- $l = 90$: for *Sanjarī zīj*: 291ʳ,15; 324ʳ,3

Geographical lists: qibla for 30 cities: 291ʳ
- πόλεις ἐπίσημοι: 338ᵛ; 458ʳ to 459ᵛ

[1] Excluding names from lists of "famous cities" and from the "seven climata"; *cf. also* μῆκος (p. 12).

PERSONAL NAMES AND WORKS QUOTED IN VAT. GR. 1058, FOL. 261 TO 459

The numbers refer to the folios[1]

[1] Not included are names which occur commonly in connection with calendars, e.g. Arabs, Persians, Romans, etc.

ARABIC AND PERSIAN TECHNICAL TERMS

awwal ἀουάλ *cf.* τασιρὴν ἀουάλ
basīṭa πασιτά
daqā'iq-i ḥiṣaṣ τακκαὲκ ἑσσάς
faḍl al-dawr cf. Appendix 16
haylāj αἰλάτζ
ḥiṣaṣ ἑσσάς *cf. daqā'iq-i ḥiṣaṣ*
ikhtilāf ἐκτλῆφι
intihā' ἰντεέ
kabīsa καπισά
kaid κάιτ
khāṣṣa χατζᾶ *cf.* χασσᾶ
khāṣṣa mu'addal χασσᾶ μαιτάλ
khusūf χουσούφ
kusūf κουσούφ
manẓar μανδάρ
maqūm μακόμ
maṭāla' ματαλέ
marfū' μορφοῦ
mu'addal μαδάλ; *cf. also khāṣṣa mu'addal*
sā'at σαάτ *cf.* ἑσσὰ σαὰτ μπότ and ντατὶλ σαὰτ ῥοέτ
ta'dīl-i sā'at-i rū'ya ντατὶλ σαὰτ ῥοέτ
tasyīr-i awwal τασιρὴν ἀουάλ
zīj ζῆζι
zubra ζουμπρᾶ

INDEX VERBORUM

Words without further references are to be found in the Glossary of Technical Terms (p. 6 ff.) at their proper alphabetic place; otherwise all, or additional, passages are listed under the words after the colon or in an appendix (p. 19 ff.).

BIBLIOGRAPHY

Almagest: Claudii Ptolemaei opera I, Syntaxis mathematica, ed. J. L. Heiberg, 2 vols., Leipzig, Teubner, 1898, 1903. Des Claudius Ptolemäus Handbuch der Astronomie, übers. von Karl Manitius, 2 vols., Leipzig, Teubner, 1912, 1913.

Bīrūnī, Astrol.: The Book of Instruction in the Elements of the Art of Astrology by . . . al-Bīrūnī, ed. and trsl. by R. R. Wright, London, Luzac, 1934.

Bouché-Leclercq, AG: A. Bouché-Leclercq, L'Astrologie grecque, Paris, 1899.

Bullialdus, Astr. Philol.: Ismaelis Bullialdi Astronomia Philolaica, Parisiis 1645 (in particular the 2nd part: Tabulae Philolaicae, Georgii medici Chrisococcae Expositio Syntaxeos Persarum [p. 211-232]).

CCAG: Catalogus Codicum Astrologorum Graecorum, 12 vols., Bruxelles 1898–1953.

Cod. Vat. Gr.: Codices Vaticani Graeci, rec. J. Mercati et P. Franchi de' Cavalieri, Roma, 1923 ff.

Delambre, HAMA: Delambre, Histoire de l'astronomie du moyen age, Paris, 1819.

Halma: Commentaire de Théon d'Alexandrie etc., 3 vols., Paris, 1822 to 1825.

Heiberg, Byz. Anal.: J. L. Heiberg, Byzantinische Analekten, Abh. zur Geschichte der Mathematik 9 (1899) [= Supplement 14 zur Zeitschrift für Mathematik und Physik 44] p. 161 to 174.

Heiberg or Heib.: *see* Almagest.

Honigmann, Ostgrenze: E. Honigmann, Die Ostgrenze des Byzantinischen Reiches von 363 bis 1071, Corpus Bruxellense Historiae Byzantinae 3, Bruxelles, 1935.

Kennedy, Survey: E. S. Kennedy, A survey of Islamic astronomical tables. *Trans. Amer. Philos. Soc.* **46** (2), 1956.

Luckey, Rechenkunst: Paul Luckey, Die Rechenkunst bei Ǧamšīd b. Mas'ūd al-Kāšī, Abh. für die Kunde des Morgenlandes 31,1, Wiesbaden, 1951.

Manitius: *see* Almagest.

Nallino, Batt.: Al-Battānī, Opus astronomicum, ed. C. A. Nallino. Vols. I and II, Milano, 1903, 1907 (= Pubbl. Osservatorio di Brera 40).

Nallino, Scritti: C. A. Nallino, Raccolta di scritti editi e inediti, Roma, Istituto per l'oriente, 1944.

Neugebauer, Ex. Sci.: O. Neugebauer, The exact sciences in antiquity, 2nd. ed., Providence, Brown University Press, 1957.

Pappus, Comm.: Commentaires de Pappus et de Théon d'Alexandrie sur l'Almageste, ed. A. Rome. Tome 1, Pappus d'Alexandrie, Commentaire sur les livres 5 et 6 de l'Almageste. Studi e Testi 54, Roma, Bibl. Apostolica Vaticana, 1931.

Ptolemy: *see* Almagest.

Schroeter, Sonnenfinsternisse: J. Fr. Schroeter, Spezieller Kanon der zentralen Sonnen- und Mondfinsternisse, welche innerhalb des Zeitraums von 600 bis 1800 n. Chr. in Europa sichtbar waren. Kristiania, 1923.

Sédillot, Tables astron.: E. A. Sédillot, Prolégomènes des tables astronomiques d' Oloug–Beg. Paris, 1853.

Setton, Byz. Backgr.: Kenneth M. Setton, The Byzantine background to the Italian renaissance, *Proc. Amer. Philos. Soc.* **100**: 1-76, 1956.

Suter, Khw.: Die astronomischen Tafeln des Muḥammed ibn Mūsā al-Khwārizmī . . . auf Grund der Vorarbeiten von A. Bjørnbo und R. Besthorn . . . herausg. von H. Suter. Danske Vidensk. Selsk. Skrifter 7, Hist. fil. Afd. III,1, Copenhagen, 1914.

Suter, MMA: H. Suter, Die Mathematiker und Astronomen der Araber und ihre Werke, Leipzig, 1900 (= Abh. z. Gesch. d. math. Wiss. 10).

Tannery, Mém. Sci.: P. Tannery, Mémoires Scientifiques, Paris, 1912 ff.

Usener, Kl. Schr. III: H. Usener, Kleine Schriften III, Leipzig, 1914.

van der Waerden, H. T.: B. L. van der Waerden, Die Handlichen Tafeln des Ptolemaios, Osiris 13, p. 54-78, 1959.

SUBJECT INDEX

The numbers refer to the pages, Greek words to the Glossary of Technical Terms (p. 6 ff.)

www.ingramcontent.com/pod-product-compliance
Lightning Source LLC
LaVergne TN
LVHW081603100826
845153LV00004B/449

9781422376508